Principles of
Environmental Chemistry

Principles of Environmental Chemistry

Jonathan Ayers

Larsen & Keller
www.larsen-keller.com

Principles of Environmental Chemistry
Jonathan Ayers
ISBN: 978-1-64172-462-3 (Hardback)

⊟ Larsen & Keller

Published by Larsen and Keller Education,
5 Penn Plaza,
19th Floor,
New York, NY 10001, USA

Cataloging-in-Publication Data

Principles of environmental chemistry / Jonathan Ayers.
   p. cm.
Includes bibliographical references and index.
ISBN 978-1-64172-462-3
1. Environmental chemistry. 2. Chemistry. 3. Green chemistry. I. Ayers, Jonathan.
TD193 .P75 2020
577.14--dc23

For more information regarding Larsen and Keller Education and its products, please visit the publisher's website www.larsen-keller.com

Table of Contents

Preface

The study of the biochemical and chemical phenomena that occur in natural places falls under the domain of environmental chemistry. It studies the sources, reactions, effects and transport of chemical species in diverse environments such as soil, air and water. It involves a better understanding of how the uncontaminated environment works. It also includes the study of chemicals in concentrations that are present naturally. It is an interdisciplinary field of science, which includes various branches of chemistry such as aquatic, soil and atmospheric chemistry. It also uses principles from analytical chemistry. Environmental chemistry observes the chemical processes occurring in the environment that are affected by human activities and causes various environmental changes. This textbook presents the complex subject of environmental chemistry in the most comprehensible and easy to understand language. Some of the diverse topics covered in this textbook address the varied branches that fall under this category. Those in search of information to further their knowledge will be greatly assisted by this book.

To facilitate a deeper understanding of the contents of this book a short introduction of every chapter is written below:

Chapter 1- Environmental chemistry deals with the study of chemical processes occurring in the environment and the effects of human activities on them. It includes astrochemistry, atmospheric chemistry, environmental modeling, geochemistry, marine chemistry and pollution remediation. The topics elaborated in this chapter will help in gaining a better perspective about environmental chemistry.

Chapter 2- Earth is divided into four major subsystems, namely lithosphere, hydrosphere, biosphere and atmosphere. Soil chemistry focuses on the study of chemical characteristics of soil. This chapter has been carefully written to provide an easy understanding of these different spheres of the Earth and their chemistry.

Chapter 3- Biogeochemical cycles refer to the pathways concerned with the cycling of chemical substances through the biotic and abiotic atmosphere of the Earth. Oxygen cycle, nitrogen cycle, sulfur cycle, phosphorus cycle, carbon cycle, etc. are some of the cycles that fall under its domain. This chapter discusses all these biogeochemical cycles in detail.

Chapter 4- The contamination of the physical and biological components of the atmosphere is referred to as environmental pollution. It can be categorized into soil pollution, air pollution and water pollution. The topics elaborated in this chapter will help in gaining a better perspective about these types of environmental pollution.

Chapter 5- Environmental problems are the unwanted changes that are caused by natural and man-made activities in the environment. It includes acid rain, ozone depletion, photochemical smog, global warming, etc. This chapter closely examines these environmental problems to provide an extensive understanding of the subject.

Chapter 6- Environmental Monitoring is the assessment, analysis, monitoring and protection of the quality of the environmental conditions by implementation and development of policies. It includes monitoring of air quality, water quality, soil quality, etc. All the aspects of environmental monitoring have been carefully analyzed in this chapter.

I owe the completion of this book to the never-ending support of my family, who supported me throughout the project.

Jonathan Ayers

Understanding Environmental Chemistry

Environmental chemistry deals with the study of chemical processes occurring in the environment and the effects of human activities on them. It includes astrochemistry, atmospheric chemistry, environmental modeling, geochemistry, marine chemistry and pollution remediation. The topics elaborated in this chapter will help in gaining a better perspective about environmental chemistry.

The Environment

Environment means everything around to a living being. Especially the circumstances of life of people or society in their life conditions. It comprises the set of natural, social and cultural values existing in a place and at a particular time, that influence in the life of the human being and in the generations to come. I.e., it is not only the space in which life develops, but it also includes living beings, objects, water, soil, air and the relations between them as well as intangibles like culture.

Concept of Environment

The environment is a system consisting of natural and artificial elements that are interrelated and which are modified by human action. It's the environment that affects the way of life of the society, including natural, social and cultural values that exist in a place and time.

Living beings, soil, water, air, physical objects made by man and the symbolic elements (as traditions, for example) make up the environment. The conservation of this is essential to the sustainable life of current generations and generations.

It could be said that environment includes factors physical (such as climate and geology), biological (human population, flora, fauna, water) and socioeconomic (employment, urbanization, social conflicts).

Referred to as ecosystem to the set consisting of all biotic factors of an area and the abiotic factors of the environment. The ecosystem is a community of living beings with interrelated life processes.

Ecology is another notion linked to the environment, since it is the discipline that studies the relationship between human beings and their environment. Ecological behavior protects environmental resources to ensure the present and future livelihood of human beings.

For example: If in the Middle a mountain builds a huge hotel, the environment will be damaged. Ecology activists should therefore oppose such projects to preserve the ecosystem.

Environmental Chemistry

Environmental chemistry refers to the occurrence, movements, and transformations of chemicals in the environment. Environmental chemistry deals with naturally occurring chemicals such as metals, other elements, organic chemicals, and bio-chemicals that are the products of biological metabolism. Environmental chemistry also deals with synthetic chemicals that have been manufactured by humans and dispersed into the environment, such as pesticides, polychlorinated biphenyls (PCBs), dioxins, furans, and many others.

The occurrence of chemicals refers to their presence and quantities in various compartments of the environment and ecosystems. For example, in a terrestrial ecosystem such as a forest, the most important compartments to consider are the mineral soil , water and air present in spaces within the soil, the above-ground atmosphere , dead biomass within the soil and lying on the ground as logs and other organic debris, and living organisms, the most abundant of which are trees. Each of these components of the forest ecosystem contains a wide variety of chemicals in some concentration, and in some amount. Chemicals move between all of these compartments, as fluxes that represent elements of nutrient and mineral cycles.

The movements of chemicals within and among compartments often involve a complex of transformations among potential molecular states. There may also be changes in physical states, such as evaporation of liquids, or crystallization of dissolved substances. The transformations of chemicals among molecular states can be illustrated by reference to the environmental cycling of sulfur. Sulfur (S) is commonly emitted to the atmosphere as the gases sulfur dioxide (SO_2) or hydrogen sulfide (H_2S), which are transformed by photochemical reactions into the negatively-charged ion, sulfate (SO_4^{-2}). The sulfate may eventually be deposited with precipitation to a terrestrial ecosystem, where it may be absorbed along with soil water by tree roots, and later used to synthesize biochemicals such as proteins and amino acids. Eventually, the plant may die and its biomass deposited to the soil surface as litter. Microorganisms can then metabolize the organic matter as a source of energy and nutrients, eventually releasing simple inorganic compounds of sulfur such as sulfate or hydrogen sulfide into the environment. Alternatively, the plant biomass may be harvested by humans and used as a fuel, with the organic sulfur being oxidized during combustion and emitted to the atmosphere as sulfur dioxide. Organic and mineral forms of sulfur also occur in fossil fuels such as petroleum and coal, and the combustion of those materials also results in an emission of sulfur dioxide to the atmosphere.

Contamination and Pollution

Contamination and pollution both refer to the presence of chemicals in the environment, but it is useful to distinguish between these two conditions. Contamination refers to the presence of one or more chemicals in concentrations higher than normally occurs in the ambient environment, but not high enough to cause biological or ecological damages. In contrast, pollution occurs when chemicals occur in the environment in concentrations high enough to cause damages to organisms. Pollution results in toxicity and ecological changes, but contamination does not cause those damages.

Chemicals that are commonly involved in pollution include the gases sulfur dioxide and ozone, diverse kinds of pesticides, elements such as arsenic , copper , mercury , nickel , and selenium, and some naturally occurring bio-chemicals. In addition, large concentrations of nutrients such as phosphate and nitrate can cause eutrophication, a type of pollution associated with excessive eco-logical productivity. Although any of these chemicals can cause pollution in certain situations, they most commonly occur in concentrations too small to cause toxicity or other ecological damages.

Modern analytical chemistry has become extremely sophisticated, and this allows trace contami-nation of potentially toxic chemicals to be measured at levels that are much smaller than what is required to cause demonstrable physiological or ecological damages.

Environmental Chemistry of the Atmosphere

Nitrogen gas (N_2) comprises about 79% of the mass of Earth's atmosphere, while 20% is oxygen (O_2), 0.9% argon (Ar), 0.035% carbon dioxide (CO_2), and the remainder composed of a variety of trace gases. The atmosphere also contains variable concentrations of water vapor, which can range from 0.01% in frigid arctic air to 5% in humid tropical air.

The atmosphere also can contain high concentrations of gases, vapors, or particulates that are po-tentially harmful to people, other animals, or vegetation, or that cause damages to buildings, art, or other materials. The most important gaseous air pollutants (listed alphabetically) are ammonia (NH_3), carbon monoxide (CO), fluoride (F, usually occurring HF), nitric oxide and nitrogen diox-ide (NO and NO_2, together known as oxides of nitrogen, or NO_x), ozone (O_3), peroxyacetyl nitrate (PAN), and sulfur dioxide (SO_2).

Vapors of elemental mercury and hydrocarbons can also be air pollutants. Particulates with tiny diameters (less than 1μm) can also be important, including dusts containing such toxic elements as arsenic, copper, lead , nickel, and vanadium, organic aerosols that are emitted as smoke during combustions (including toxins known as polycyclic aromatic hydrocarbons), and non-reactive minerals such as silicates.

Some so-called "trace toxics" also occur in the atmosphere in extremely small concentrations. The trace toxics include persistent organochlorine chemicals such as the pesticides DDT and dieldrin, polychlorinated biphenyls (PCBs), and the dioxin, TCDD. Other, less persistent pesticides may also be air pollutants close to places where they are used.

Environmental Chemistry of Water

Earth's surface waters vary enormously in their concentrations of dissolved and suspended chemi-cals. Other than the water, the chemistry of oceanic water is dominated by sodium chloride (NaCl), which has a typical concentration of about 3.5% or 35 g/l. Also important are sulfate (2.7 g/l), magnesium (1.3 g/l), and potassium and calcium (both 0.4 g/l). Some saline lakes can have much larger concentrations of dissolved ions, such as Great Salt Lake in Utah, which contains more than 20% salts.

Fresh waters are much more dilute in ions, although the concentrations are variable among water-bodies. The most important cations in typical fresh waters are calcium (Ca^{2+}), magnesium (Mg^{2+}), sodium (Na^+), ammonium (NH_4^+), and hydrogen ion (H^+; this is only present in acidic waters, otherwise hydroxy ion or OH^- occurs). The most important anions are bicarbonate (HCO_3^-) sulfate

(SO_4^{2+}), chloride (Cl^-), and nitrate (NO_3). Some fresh waters have high concentrations of dissolved organic compounds, known as humic substances, which can stain the water a tea-like color. Typical concentrations of major ions in fresh water are: calcium 15 mg/l, sulfate 11 mg/l, chloride 7 mg/l, silica 7 mg/l, sodium 6 mg/l, magnesium 4 mg/l, and potassium 3 mg/l.

The water of clean precipitation is considerably more dilute than that of surface waters such as lakes. For example, precipitation at a remote place in Nova Scotia contained 1.6 mg/l of sulfate, 1.3 mg/l chloride, 0.8 mg/l sodium, 0.7 mg/l nitrate, 0.13 mg/l calcium, 0.08 mg/l ammonium, 0.08 mg/l magnesium, and 0.08 mg/l potassium. Because that site is about 31 mi (50 km) from the Atlantic Ocean, its precipitation is influenced by sodium and chloride originating with sea spray. In comparison, a more central location in North America had a sodium concentration of 0.09 mg/l and chloride 0.15 mg/l.

Pollution of surface waters is most often associated with the dumping of human or industrial sewage, nutrient inputs from agriculture, acidification caused by acidic precipitation or by acid-mine drainage , and industrial inputs of toxic chemicals. Eutrophication is caused when nutrient inputs cause large increases in aquatic productivity, especially in fresh waters and shallow marine waters into which sewage is dumped or that receive runoff containing agricultural fertilizers. In general, marine ecosystems become eutrophic when they are fertilized with nitrate, and freshwater systems with phosphate. Only 35–100 µg/l or more of phosphate is enough to significantly increase the productivity of most shallow lakes, compared with the background concentration of about 10 µg/l or less.

Freshwater ecosystems can become acidified by receiving drainage from bogs, by the deposition of acidifying substances from the atmosphere (such as acidic rain), and by acid-mine drainage. Atmospheric depositions have caused a widespread acidification of surface waters in eastern North America, Scandinavia, and other places. Surface waters acidified by atmospheric depositions commonly develop pHs of about 4.5–5.5. Tens of thousands of lake and running-water ecosystems have been damaged in this way. Acidification has many biological consequences, including toxicity caused to many species of plants and animals, including fish.

Some industries emit metals to the environment, and these may pollute fresh and marine waters. For instance, lakes near large smelters at Sudbury, Ontario, have been polluted by sulfuric acid, copper, nickel, and other metals, which in some cases occur in concentrations large enough to cause toxicity to aquatic plants and animals.

Mercury contamination of fish is also a significant problem in many aquatic environments. This phenomenon is significant in almost all large fish and sharks , which accumulate mercury progressively during their lives and commonly have residues in their flesh that exceed 0.5 ppm (this is the criterion set by the World Health Organization for the maximum concentration of mercury in fish intended for human consumption). It is likely, however, that the oceanic mercury is natural in origin, and not associated with human activities. Many fresh-water fish also develop high concentrations of mercury in their flesh, also commonly exceeding the 0.5 ppm criterion. This phenomenon has been demonstrated in many remote lakes. The source of mercury may be mostly natural, or it may originate with industrial sources whose emissions are transported over a long distance in the atmosphere before they are deposited to the surface. Severe mercury pollution has also occurred near certain factories, such as chlor-alkali plants and pulp mills. The most famous example occurred at Minamata, Japan, where industrial discharges led to the pollution of marine organisms, and then resulted in the poisoning of fish-eating animals and people.

Environmental Chemistry of Soil and Rocks

The most abundant elements in typical soils and rocks are oxygen (47%), silicon (28%), aluminum (8%), and iron (3–4%). Virtually all of the other stable elements are also present in soil and rocks, and all of these can occur in a great variety of molecular forms and minerals. Under certain circumstances, some of these chemicals can occur in relatively high concentrations, sometimes causing ecological damages.

This can occur naturally, as in the case of soils influenced by so-called serpentine minerals, which can contain hundreds to thousands of ppm of nickel. In addition, industrial emissions of metals from smelters have caused severe pollution. Soils near Sudbury, for example, can contain nickel and copper concentrations up to 5,000 ppm each. Even urban environments can be severely contaminated by certain metals. Soils collected near urban factories for recycling old automobile batteries can contain lead in concentrations in the percent range, while the edges of roads can contain thousands of ppm of lead emitted through the use of leaded gasoline.

Trace Toxics

Some chemicals occur in minute concentrations in water and other components of the environment, yet still manage to cause significant damages. These chemicals are sometimes referred to as trace toxics. The best examples are the numerous compounds known as halogenated hydrocarbons, particularly chlorinated hydrocarbons such as the insecticides DDT, DDD, and dieldrin, the dielectric fluids PCBs, and the chlorinated dioxin, TCDD. These chemicals are not easily degraded by either ultraviolet radiation or by metabolic reactions, so they are persistent in the environment. In addition, chlorinated hydrocarbons are virtually insoluble in water, but are highly soluble in lipids such as fats and oils. Because most lipids in ecosystems occur within the bodies of organisms, chlorinated hydrocarbons have a marked tendency to bioaccumulate (i.e., to occur preferentially in organisms rather than in the non-living environment). This, coupled with the persistence of these chemicals, results in their strong tendency to food-chain/web accumulate or biomagnify (i.e., to occur in their largest concentrations in top predators).

Fish-eating birds are examples of top predators that have been poisoned by exposure to chlorinated hydrocarbons in the environment. Some examples of species that have been affected by this type of ecotoxicity include the peregrine falcon (*Falco peregrinus*), bald eagle (*Haliaeetus leucocephalus*), osprey (*Pandion haliaetus*), brown pelican (*Pelecanus occidentalis*), double-crested cormorant (*Phalacrocorax auritus*), and western grebe (*Aechmophorus occidentalis*). Concentrations of chlorinated hydrocarbons in the water of aquatic habitats of these birds is generally less than 1 µg/l (part per billion, or ppb), and less than 1 ng/l (part per trillion, or ppt) in the case of TCDD. However, some of the chlorinated hydrocarbons can biomagnify to tens to hundreds of mg/kg (ppm) in the fatty tissues of fish-eating birds. This can cause severe toxicity, characterized by reproductive failures, and even the deaths of adult birds, both of which can cause populations to collapse.

Other trace toxics also cause ecological damages. For example, although it is only moderately persistent in aquatic environments, the insecticide carbofuran can accumulate in acidic standing water in recently treated fields. If geese, ducks, or other birds or mammals utilize those temporary aquatic habitats, they can be killed by the carbofuran residues. Large numbers of wildlife have been killed this way in North America.

Petroleum

Water pollution can also result from the occurrence of hydrocarbons in large concentrations, especially after spills of crude oil or its refined products. Oil pollution can result from accidental spills of petroleum from wrecked tankers, offshore drilling platforms, broken pipelines, and from spills during warfare, as occurred during the Gulf War of 1991. Other important sources of oil pollution include operational discharges from tankers disposing oily bilge waters, and chronic releases from oil refineries and urban runoff.

The concentration of natural hydrocarbons in seawater is about 1 ppb, mostly due to releases from phytoplankton and bacteria. Beneath a slick of petroleum spilled at sea, however, the concentration of dissolved hydrocarbons can exceed several ppm, enough to cause toxicity to some organisms. There are also finely suspended droplets of petroleum in water beneath slicks, as a result of wave action on the floating oil. The slick and the sub-surface emulsion of oilin-water are highly damaging to organisms that become coated with these substances.

- Contaminant: A substance present in greater than the usual (normal) concentration.

- Pollutant: A contaminant, which has a detrimental effect upon its environment (or something of value within it). A pollutant can be anthropogenic (human made) or biogenic (from a natural source).

- Source: Every pollutant originates from a source. Identifying the source is very important since the most efficient mitigation can usually be achieved at this location. Point sources are often easy to identify as they are associated with a specific industry or activity. Non-point sources give rise to a more diffuse pollutant loading and can be more difficult to identify and remediate.

- Receptor: Anything that a pollutant acts upon.

- Reservoir: A set of environmental conditions in which a substance resides and usually mixes. The atmosphere, oceans and other water bodies are reservoirs for many substances. When the concentration of a substance within a reservoir does not appear to change, the rate of inflow = rate of outflow and the situation is said to have reached a steady-state concentration.

- Sink: A longer term reservoir in which a substance is essentially immobilized. Such a repository may be natural or human-made. The oceans and ocean sediments are a sink for many of the dissolved species present in freshwater.

- Residence time (also known as the lifetime): Is the length of time for the amount of a substance to decrease to 1/e (i.e., 0.37) of it's original value, if the input were to suddenly cease. It can be defined as the amount of material in a reservoir divided by it's total rate of inflow (or outflow).

$$\text{Residence time} = \frac{\text{amount of substance in the reservoir}}{\text{rate of inflow to (or outflow from) reservoir}}$$

Note that the units must be chosen to be compatible, so that they cancel and the residence time has units of time. For example, the numerator of this expression may be expressed in grams, moles per liter, m^3 or tonnes. Therefore, the denominator must have corresponding units, such as, grams per second, moles per liter per minute, m^3 per day or tonnes per year.

- Chemical Speciation: The chemical form of a given element under a set of environmental conditions (i.e., temperature, pressure, pH, pe). Depends on a variety of physical and chemical characteristics of the species involved.

Chemistry of each Sphere

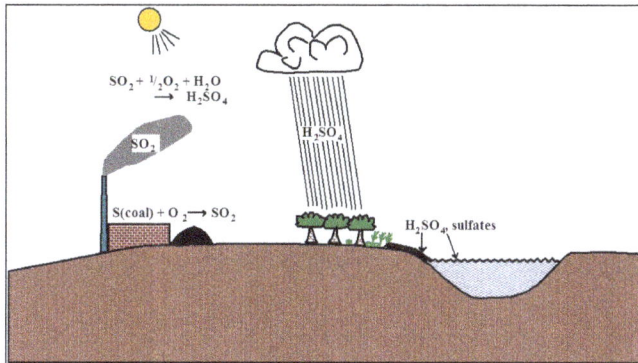

Aquatic Chemistry.

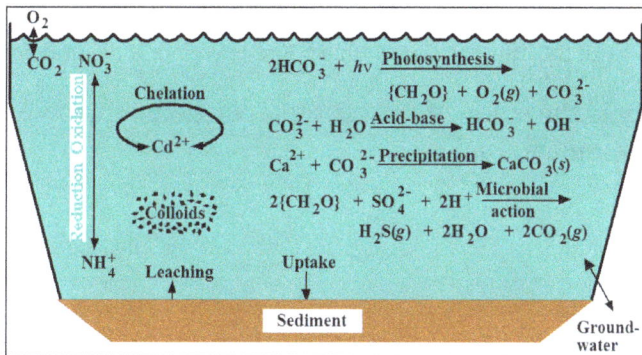

Atmospheric Chemistry.

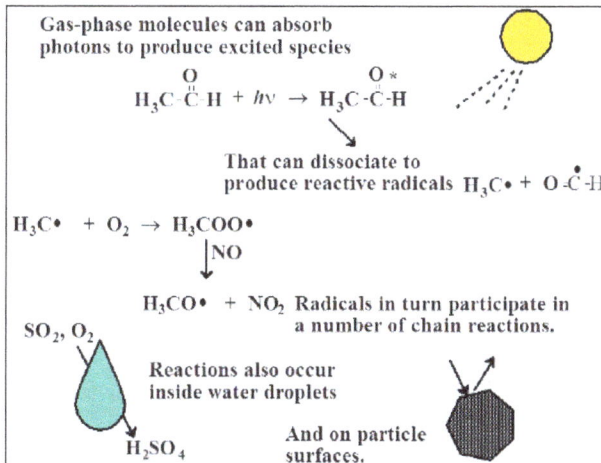

Chemistry of the Geosphere and Soil.

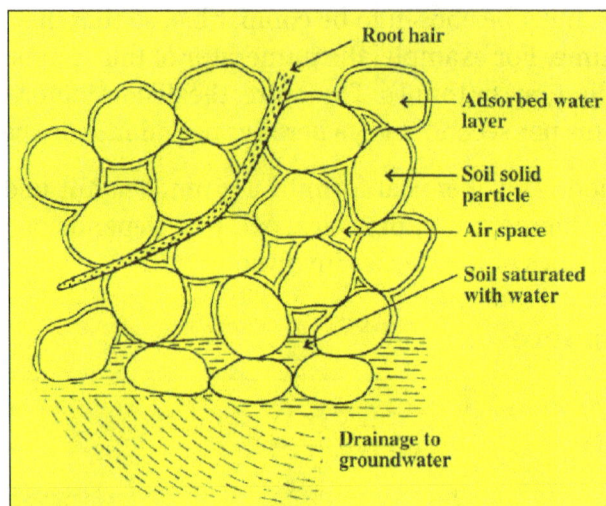

Chemistry of the Biosphere and Toxicological Chemistry.

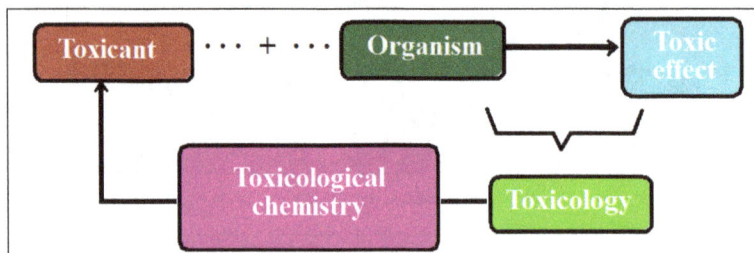

Chemistry of the Anthrosphere within a Framework of Industrial Ecology.

Environmental chemistry is the study of the sources, reactions, transport, and fates of chemical species involving all environmental spheres.

Earth's Spheres and their Chemistry

Earth is divided into four major subsystems, namely lithosphere, hydrosphere, biosphere and atmosphere. Soil chemistry focuses on the study of chemical characteristics of soil. This chapter has been carefully written to provide an easy understanding of these different spheres of the Earth and their chemistry.

Atmosphere and its Chemistry

One of the four components of the Earth's ecosystem (the other three are biosphere, hydrosphere, and lithosphere), it is a band of gases enveloping the Earth's surface. Ninety-nine percent of its mass is concentrated within 20 miles of the earth's surface, and its two largest constituents (in the lowest part) are nitrogen (about 78 percent) and oxygen (about 21 percent). The remaining one percent includes mostly argon and minute amounts of carbon dioxide, helium, krypton, methane, neon, sulfur dioxide, water vapor, and xenon. In meteorology, atmosphere is divided into seven layers called regions (from the lowest to the highest): troposphere, stratosphere, mesosphere, chemosphere, thermosphere, ionosphere, and exosphere.

Atmospheric chemistry involves study of the chemistry of the atmospheres of Earth and other planets. It is a branch of atmospheric science and is a multidisciplinary field of research, drawing on environmental chemistry, meteorology, physics, computer modeling, oceanoraphy, geology, volcanology, and other disciplines. In addition, it is being increasingly associated with the field known as climatology.

The atmosphere consists of 78% nitrogen, 21% oxygen, 1% water vapor, and a minute amount of other trace gases like argon, and carbon monoxide. All of these gases combine to absorb ultraviolet radiation from the Sun and warm the planet's surface through heat retention. The mass of the atmosphere is around 5×10^{18} kg. 75% of the atmospheric mass is within 11 km of the surface. While the atmosphere becomes thinner the higher you go, there is no clear line demarcating the atmosphere from space; however, the Karman line, at 100 km, is often regarded as the boundary between atmosphere and outer space. The effects of re-entry can be felt at 120 km.

Over the vast history of Earth there have been three different atmospheres or one that has evolved in three major stages. The first atmosphere came into being as a result of a major rainfall over the entire planet that caused the build up of a major ocean. The second atmosphere began to develop around 2.7 billion years ago. The presence oxygen began to appear apparently from being released by photosynthesizing algae. The third atmosphere came into play when the planet began to stretch its legs, so to speak. Plate tectonics began constantly rearranging the continents about 3.5 billion years ago and helped to shape long-term climate evolution by allowing the transfer of carbon dioxide to large land-based carbonate stores. Free oxygen did not exist until about 1.7 billion years ago and this can be seen with the development of the red beds and the end of the banded iron formations. This signifies a shift from a reducing atmosphere to an oxidizing atmosphere. Oxygen showed major ups and downs until reaching a steady state of more than 15%.

The Earth's atmosphere performs a couple of cool optical tricks. The blue color of the sky is due to Rayleigh scattering which means as light moves through the atmosphere, most of the longer wavelengths pass straight through. Very little of the red, orange and yellow light is affected by the air; however, much of the shorter wavelength light(blue) is absorbed by the gas molecules. The absorbed blue light is then radiated in every direction. So, no matter where you look, you see the scattered blue light. The atmosphere is also responsible for the aurora borealis. Auroras are caused by the bombardment of solar electrons on oxygen and nitrogen atoms in the atmosphere. The electrons literally excite the oxygen and nitrogen atoms high in the atmosphere to create the beautiful light show we know as an aurora.

The atmosphere is divided into 5 major zones. The troposphere begins at the surface and extends to between 7 km at the poles and 17 km at the equator, with some variation due to weather. The stratosphere extends to about 51 km. The mesosphere extends to about 85 km. Most meteors burn up in this zone of the atmosphere. The thermosphere extends up to between 320 and 380 km. This is where the International Space Station orbits. The temperature here can rise to 1,500 °C. The exosphere is the last bastion of the atmosphere. Here the particles are so far apart that they can travel hundreds of km without colliding with one another. The exosphere is mainly composed of hydrogen and helium.

Atmospheric Layers

The properties of the Earth's atmosphere vary with altitude. Based on these properties, the atmosphere may be regarded as having different layers or zones. According to one system of nomenclature, there are five layers: the *troposphere, stratosphere, mesosphere, thermosphere,* and *exosphere.* The boundaries between these regions are called the *tropopause, stratopause, mesopause,* and *exobase.*

Troposphere

The troposphere is the atmosphere's lowest and densest layer, and it is also known as the lower atmosphere. It starts from the Earth's surface and reaches up to about 7 km (4.35 miles) at the poles and 17 km (10.5 miles) at the equator, with some variation caused by weather factors. The upper boundary of this layer is called the tropopause.

The troposphere has a great deal of vertical mixing of its constituent gases, as a result of solar heat. As the Sun heats the Earth's surface, the air masses near the surface become warmer, rise, and release heat. That process further buoys the air masses. This sequence continues until all the water vapor in the air is removed.

The temperature of the troposphere decreases with height. At middle latitudes, the temperature drops from about +17° C (62° F)at sea level to about -52° C (-62° F) at the beginning of the tropopause. At the poles, the troposphere is thinner and the temperature falls to only -45° C (-49° F), while at the equator, the temperature at the top of the troposphere can reach -75° C (-103° F).

Stratosphere

The stratosphere is situated directly above the troposphere and just below the mesosphere. In terms of its altitude range, it lies between about 10 km and 50 km at moderate latitudes, but it starts at about 8 km at the poles. This layer is dynamically stable, with no regular mixing of air and associated turbulence.

The stratosphere is stratified (layered) in temperature, with warmer layers higher up and cooler layers farther down—the reverse of the temperature variation in the troposphere. The upper layers of the stratosphere are heated by the presence of an ozone layer that absorbs ultraviolet (UV) radiation from the Sun—radiation that would otherwise be harmful to living organisms on Earth. The stratopause (at the top of the stratosphere) has a temperature of about 270K—fairly close to the temperature at ground level.

The base of the stratosphere occurs where heating by conduction from above and convection from below (through the troposphere) balance out. The stratosphere begins at lower altitudes near the poles because the ground temperatures are lower there.

Commercial airliners typically cruise at an altitude near 10 km in temperate latitudes, in the lower reaches of the stratosphere. In this manner, they avoid the atmospheric turbulence that occurs in the troposphere.

Mesosphere

The mesosphere is the layer between about 50 km and about 80–85 km above the Earth's surface. It is sandwiched between the stratosphere and the thermosphere. The temperature in this layer decreases with increasing altitude and can be as low as 200K ($\approx$ -73° C, -99° F), varying according to latitude and season.

Given that it lies between the maximum altitude for most aircraft and the minimum altitude for most spacecraft, this region of the atmosphere is directly accessible only through the use of sounding rockets. As a result, it is one of the most poorly understood regions of the atmosphere.

Millions of meteors burn up daily in the mesosphere, as a result of collisions with the gas particles contained there, leading to a high concentration of iron and other metal atoms. The collisions almost always create enough heat to burn the falling objects long before they reach the ground. Thus the mesosphere protects the Earth from a barrage of would-be meteorites.

The stratosphere and mesosphere are referred to as the middle atmosphere. Regions above the mesosphere are called the upper atmosphere.

The mesopause, at an altitude of about 80 km, separates the mesosphere from the thermosphere. The mesopause lies near the turbopause, the band below which different chemical species are well mixed by turbulent eddies.

Thermosphere

The thermosphere extends from an altitude of 80–85 km to 640+ km. It lies directly above the mesosphere and right below the exosphere.

At these high altitudes, the residual atmospheric gases sort into strata according to their molecular masses. The temperature in this layer increases with altitude, due to the absorption of extremely energetic solar radiation by the small amount of oxygen present. Temperatures are highly dependent on solar activity and can rise to 2,000° C. Solar radiation causes the air particles in this layer to become ionized, that is, electrically charged.

Astronauts travel at altitudes that exceed 80 km (50 miles), which means that they travel within or go beyond the thermosphere. An altitude of 120 km (75 miles) marks the boundary where atmospheric effects become noticeable during re-entry.

Exosphere

The exosphere is the uppermost layer of the atmosphere. Its lower boundary at the edge of the thermosphere is estimated to be 500 km to 1,000 km above the Earth's surface, and its upper boundary at about 10,000 km.

It is only from the exosphere that atmospheric gases can, to any appreciable extent, escape into outer space. The main gases in the exosphere are the lightest ones, mainly hydrogen and helium, with some atomic oxygen near the exobase (the lowest altitude of the exosphere). The few particles of gas here can reach 2,500° C (4,500° F) during the day.

The atmosphere in this layer is sufficiently rarified for satellites to orbit the Earth, although they still receive some atmospheric drag. The exact altitude at which the exosphere ends and space begins is not well-defined, and attempting to attach a specific value to it is not particularly useful.

Additional Atmospheric Regions

Atmospheric regions are also named in other ways, as follows:

- Ozone layer (Ozonosphere): In the stratosphere, in an altitude range of about 10–50 km, the concentration of ozone (O_3) is a few parts per million, which is much higher than the ozone concentration in the lower atmosphere (although it is still small compared to the main components of the atmosphere). This layer, known as the ozone layer, is vitally important to life because it absorbs biologically harmful UV radiation from the Sun. Moreover, the absorbed solar energy raises the temperature of this part of the atmosphere, creating a thermal barrier that helps trap the atmosphere below, preventing it from bleeding out into space.

- Ionosphere: This is the region of the atmosphere that contains ions (that form a "plasma"), created by the interaction of solar radiation with gas particles. The ionosphere overlaps with the mesosphere and thermosphere, going up to an altitude of 550 km. Its value in practical terms is that it enables the propagation of radio wave signals, which bounce off the ions and can be transmitted to distant places on the Earth.

- Magnetosphere: It is the region where the Earth's magnetic field interacts with the solar wind. Its inner boundary is the ionosphere, but it extends for tens of thousands of kilometers, with a long tail away from the Sun.

- Van Allen radiation belts: These are regions where charged particles (forming a plasma) from the solar wind are trapped by the Earth's magnetic field. When the belts "overload," particles strike the upper atmosphere and fluoresce, producing the effects known as the polar auroras. Qualitatively, there are two belts: an inner belt, consisting mostly of protons, and an outer belt, consisting mostly of electrons.

- Homosphere (or Turbosphere) and Heterosphere: The region below the turbopause (that is, below an altitude of about 100 km) is known as the homosphere or turbosphere, where the chemical constituents are well mixed and the composition of the atmosphere remains fairly uniform. The region above the turbopause is called the heterosphere, where, in the absence of mixing, the chemical composition of the atmosphere varies.

Pressure, Density and Mass

- Atmospheric pressure (or barometric pressure) is a direct result of the weight of the air. It is highest at the Earth's surface and decreases with altitude. This is because air at the surface is compressed by the weight of all the air above it. Air pressure varies with location and time, because the amount (and weight) of air above the Earth varies with location and time.

- Atmospheric pressure drops by approximately 50 percent at an altitude of about 5 km. (In other words, about 50 percent of the total atmospheric mass is within the lowest 5 km). The average atmospheric pressure at sea level is about 101.3 kilopascals (about 14.7 pounds per square inch).

- The density of air at sea level is about 1.2 kg/m^3, and it decreases as altitude increases.

- The average mass of the atmosphere is about 5,000 trillion metric tons.

Thickness of the Atmosphere

- 57.8 percent of the atmosphere is below the summit of Mount Everest.

- 72 percent of the atmosphere is below the common cruising altitude of commercial airliners (about 10,000 m or 32,800 ft).

- 99.99999 percent of the atmosphere is below the highest flight altitude of the aircraft X-15, which reached 354,300 ft (108 km) on August 22, 1963. Therefore, most of the atmosphere (99.9999 percent) is below 100 km, although in the rarified region above this there are auroras and other atmospheric effects.

- The atmosphere exists at altitudes of 1,000 km and higher, but it is so thin as to be considered non-existent.

Composition of the Atmosphere

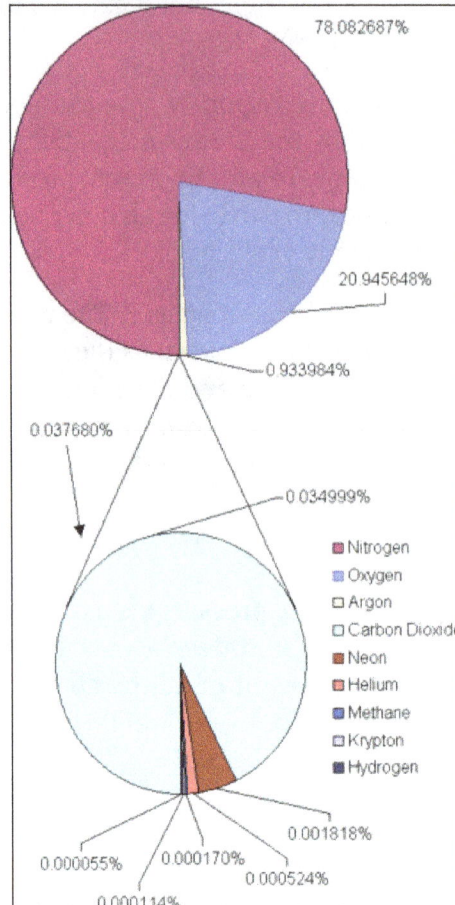

78.082687%

20.945648%

0.933984%

0.037680%

0.034999%

- Nitrogen
- Oxygen
- Argon
- Carbon Dioxide
- Neon
- Helium
- Methane
- Krypton
- Hydrogen

0.001818%

0.000055%   0.000170%

0.000524%

0.000114%

Composition of Dry Atmosphere by Volume

ppmv: parts per million by volume	
Gas	Volume
Nitrogen (N_2)	780,840 ppmv (78.084%)
Oxygen (O_2)	209,460 ppmv (20.946%)
Argon (Ar)	9,340 ppmv (0.9340%)
Carbon dioxide (CO_2)	350 ppmv
Neon (Ne)	18.18 ppmv
Helium (He)	5.24 ppmv
Methane (CH_4)	1.745 ppmv
Krypton (Kr)	1.14 ppmv
Hydrogen (H_2)	0.55 ppmv
Not included in above dry atmosphere:	
Water vapor (highly variable)	typically 1%

Gas	Volume
nitrous oxide	0.5 ppmv
xenon	0.09 ppmv
ozone	0.0 to 0.07 ppmv
nitrogen dioxide	0.02 ppmv
iodine	0.01 ppmv
carbon monoxide	trace
ammonia	trace

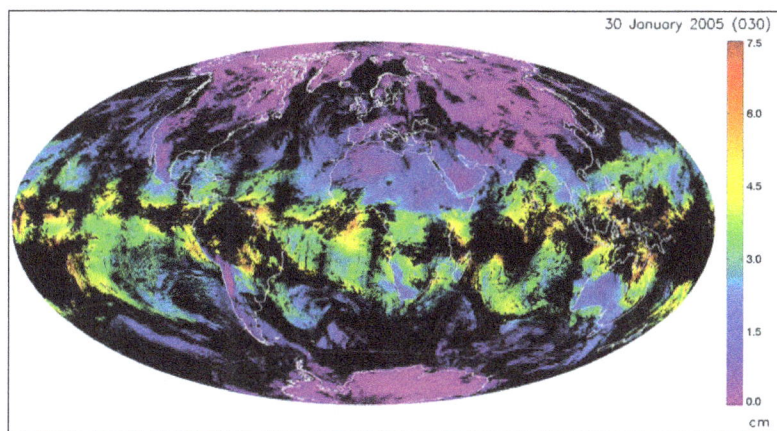

Mean Atmospheric Water Vapor.

Composition of the Heterosphere

Above the turbopause (about 100 km), the Earth's atmosphere begins to have a composition that varies with altitude. This is essentially because, in the absence of mixing, the density of a gas falls off exponentially with increasing altitude, at a rate that depends on the molecular mass of the gas. Higher mass constituents, such as oxygen and nitrogen, fall off more quickly than lighter constituents such as helium, molecular hydrogen, and atomic hydrogen. Thus, as the altitude increases in the heterosphere, the atmosphere is dominated successively by helium, molecular hydrogen, and atomic hydrogen. The precise altitude of the heterosphere and the layers it contains varies significantly with temperature.

Biological Significance

The Earth's atmosphere plays a vital role in sustaining life on this planet. Oxygen is needed for respiration by animals, plants, and some bacteria. Nitrogen is an inert gas that reduces the amount of oxygen available for the oxidation of natural materials, thus restricting spontaneous combustion (burning) of flammable materials and the corrosion of metals. Nitrogen is also used by "nitrogen-fixing" bacteria to produce compounds that are useful for plant growth. Plants that perform photosynthesis take up carbon dioxide from the air and release oxygen. Carbon dioxide and water vapor act as "greenhouse gases" that keep the Earth sufficiently warm to maintain life. Water vapor in the air is part of the water cycle that produces precipitation (such as rain and snow) that replenishes moisture in the soil. In addition, water vapor prevents exposed living tissue from drying up.

Moreover, several regions of the atmosphere exert their protective effect from a distance. For instance, the ozone layer absorbs UV radiation that can damage the tissues and genetic material of living organisms. The mesosphere, in which millions of meteors burn up daily, protects the Earth's surface from being continually bombarded by these falling objects. The magnetosphere, which extends well beyond the atmosphere, protects the Earth from the damaging rain of charged particles carried by the solar wind.

Chemical Composition of Air

Air is a mixture of gases which makes up the Earth's atmosphere. These gases are colourless and odourless and hence, we can't see them but only feel them. The atmosphere is an ocean of these gases. It consists of 78% nitrogen, 21% oxygen and 1 % other gases and water vapour. The moisture content varies from place to place. Arid regions have less moisture content as compared to wetlands.

- The water vapour or moisture content of air varies. The maximum moisture carrying capacity of air depends primarily on temperature.

- The composition of air is unchanged until the elevation of approximately 10.000 m.

- The average air temperature diminishes at the rate of 0.6°C for each 100 m vertical height.

- "One Standard Atmosphere" is defined as the pressure which is equal to that exerted by a 760 mm column of mercury at 0°C sea level and at standard gravity (32.174 ft/sec2).

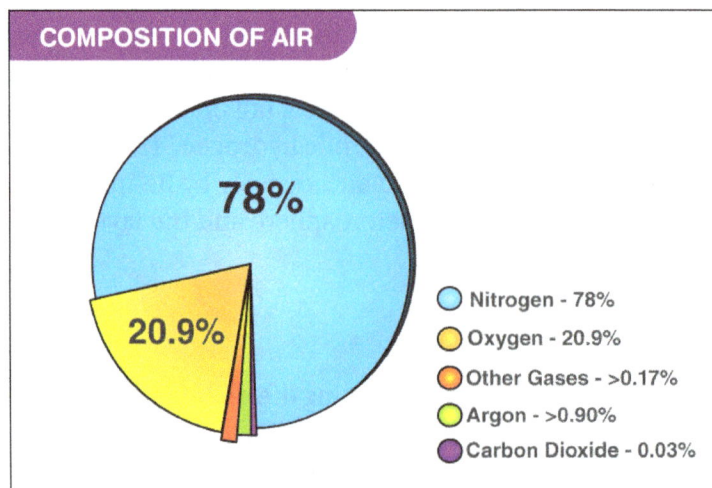

COMPOSITION OF AIR

78%

20.9%

- Nitrogen - 78%
- Oxygen - 20.9%
- Other Gases - >0.17%
- Argon - >0.90%
- Carbon Dioxide - 0.03%

Composition of Air				
Element	Volume by %	Weight by %	PPM(Parts per Million) by Volume	Symbol of the Element	Molecular Weight of the element
Nitrogen	78.08	75.47	780790	N_2	28.01
Oxygen	20.95	23.20	209445	O_2	32.00
Argon	0.93	1.28	9339	Ar	39.95

Carbon Dioxide	0.040	0.062	404	CO_2	44.01
Neon	0.0018	0.0012	18.21	Ne	20.18
Helium	0.0005	0.00007	5.24	He	4.00
Krypton	0.0001	0.0003	1.14	Kr	83.80
Hydro-gen	0.00005	Negligible	0.50	H_2	2.02
Xenon	8.7×10^{-6}	0.00004	0.087	Xe	131.30

Other Components of Air

Some other components of air are mentioned below:

- Sulfur dioxide(SO_2) – 1.0 ppm.

- Methane(CH_4) – 2.0 ppm.

- Nitrous oxide(N_2O) – 0.5 ppm.

- Ozone(O_3) – 0 to 0.07 ppm.

- Nitrogen dioxide(NO_2) – 0.02 ppm.

- Iodine(I_2) – 0.01 ppm.

- Carbon monoxide(CO) – 0 to trace ppm.

- Ammonia(NH_3) – 0 to trace ppm.

Properties of Air

Gases are matter. There are certain properties of gases like any other matter. Some common properties are as follows:

- Colourless and odourless: Air generally has no colour or odour. It is an invisible matter that can only be felt. All living things breathe air for their survival. Moving air is called wind.

- Occupy space: It is a mixture of different gases. Hence, like every other matter, they also occupy space. On blowing, a balloon expands because the air being blown into it fills up the empty space.

- Air exerts pressure: It has weight, and the pressure exerted by the weight of air is known as air pressure. Due to gravity, this mixture of gases near the surface is denser than at high altitudes. This is why the gaseous atmosphere in the mountains is thinner than that at the surface.

- Expansion: Another property is its expanding property. On heating, it expands and occupies more space. The more it expands, the thinner it becomes. Hence, the pressure of the warm wind is lower than that of cold wind.

Oxygen

The most important gas in the composition is oxygen. Oxygen is an important gas, both for humans as well as the atmosphere. It supports in breathing for living beings, but at the same time, oxygen is a combustible gas, which means, it can catch fire quickly. Thus, when you light a candle, it easily lights up. Try doing the same in a vacuum, and you will not be able to light it. Which means air consists of combustible substance too. This is why the composition of oxygen in the air is just about 21%. Just enough to give us healthy lungs, but not enough to start a random fire.

Nitrogen

To balance out oxygen, there is Nitrogen. It is a non-combustible gas and occupies the highest percentage of air. This is why, when you blow on a candle it is easily extinguished. This is done because of the presence of nitrogen. Nitrogen is also one of the most abundantly found gases on earth.

Argon

Presence of Argon in the air was first suspected by Henry Cavendish in 1785, but it was not confirmed until 1894 when Lord Rayleigh and Sir William Ramsay attested to it. This gas is known to occur naturally in the environment and easily dissipates in a well-ventilated space. In itself, argon is an asphyxiant and inhalation of it can lead to dizziness and nausea in a person.

Carbon Dioxide

Carbon dioxide is infused in the air due to respiration. Living beings inhale oxygen and exhale carbon dioxide. There are other ways by which carbon dioxide is infused in the atmosphere, and its composition keeps changing depending on the place where you are. In general, it should occupy about 0.38% of the earth's total atmosphere.

Water Vapor

Finally, we have water vapor. When the water present in the water bodies evaporates due to heat, it rises up and mixes into the atmosphere. You can assess the presence of water vapor in the air by measuring the humidity level. You will notice that as we come closer to sea, the level of humidity rises. This is because the amount of water vapor is more near seaside areas.

Other Particles

Apart from the natural gases, air also contains things like smoke and dust. The closer you are to the city, the amount of these particles in the air increases.

Hydrosphere

The hydrosphere is the component of the Earth that is composed of all liquid water found on the planet. The hydrosphere includes water storage areas such as oceans, seas, lakes, ponds, rivers, and streams. Overall, the hydrosphere is very large, with the oceans alone covering about 71% of the surface area of Earth.

The spring shown above would be considered part of the hydrosphere.

The motion of the hydrosphere and the exchange of water between the hydrosphere and cryosphere is the basis of the hydrologic cycle. The continuous movement and exchange of water helps to form currents that move warm water from the tropics to the poles and help regulate the temperature of the Earth. The exchanging of water is thus a vital part of the hydrosphere.

It is important to note that although the hydrosphere is primarily composed of water, there are also some "impurities" or additions to this water that include dissolved minerals, dissolved gases, and particulates. Some of these can be considered pollution, while others are necessary for health of ecosystems. For example, too much sediment is harmful to the surrounding ecosystems, while insufficient levels of dissolved oxygen in the water lead to hypoxic conditions that can harm ecosystems. Thus a delicate balance is needed for healthy ecosystems that surround different components of the hydrosphere.

Components

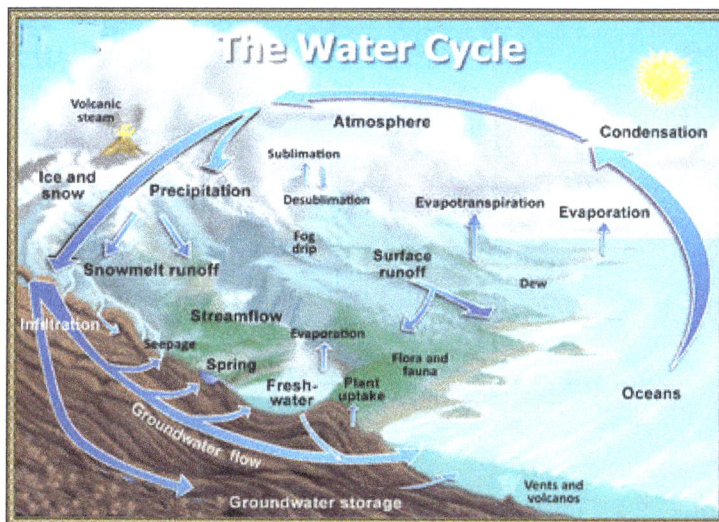
The hydrosphere cycle.

Any water storage area on the Earth that holds liquid water is considered to be a part of the hydrosphere. Because of this, there is an extensive list of formations that make up the hydrosphere. These include:

- Oceans: Most of the water on the planet Earth is salt water, and the vast majority of this salt water is held in the oceans.

- Fresh water: Fresh water is much less abundant than salt water, and is held in a variety of different places:

    ○ Surface water: Surface sources of freshwater include lakes, rivers, and streams.

    ○ Ground water: Fresh water held beneath ground makes up a small portion of the fresh water on Earth.

- Glacial water: Water that melts off of glaciers.

- Atmospheric water vapour.

Distribution and Quantity of Earth's Waters

Ocean waters and waters trapped in the pore spaces of sediments make up most of the present-day hydrosphere. The total mass of water in the oceans equals about 50 percent of the mass of sedimentary rocks now in existence and about 5 percent of the mass of Earth's crust as a whole. Deep and shallow groundwaters constitute a small percentage of the total water locked in the pores of sedimentary rocks—on the order of 3 to 15 percent. The amount of water in the atmosphere at any one time is trivial, equivalent to roughly 13,000 cubic km (about 3,100 cubic miles) of liquid water, or about 0.001 percent of the total at Earth's surface. This water, however, plays an important role in the water cycle.

Water masses at Earth's surface		
reservoir	volume (in cubic kilometres)	percent of total
oceans	1,338,000,000	96.5
ice caps, glaciers, and permanent snow	24,064,000	1.74
ground ice and permafrost	300,000	0.22
groundwater (total)	23,400,000	1.69
groundwater (fresh)	10,530,000	0.76
groundwater (saline)	12,870,000	0.93
lakes (total)	176,400	0.013
lakes (fresh)	91,000	0.007
lakes (saline)	85,400	0.006
soil moisture	16,500	0.001
atmosphere*	12,900	0.001
swamp water	11,470	0.0008
rivers	2,120	0.0002
biota	1,120	0.0001
total**	1,409,560,910	101.67

At present, ice locks up a little more than 2 percent of Earth's water and may have accounted for as much as 3 percent or more during the height of the glaciations of the Pleistocene Epoch (2.6 million to 11,700 years ago). Although water storage in rivers, lakes, and the atmosphere is small, the rate of water circulation through the rain-river-ocean-atmosphere system is relatively rapid. The amount of water discharged each year into the oceans from the land is approximately equal to the total mass of water stored at any instant in rivers and lakes.

Soil moisture accounts for only 0.005 percent of the water at Earth's surface. It is this small amount of water, however, that exerts the most direct influence on evaporation from soils. The biosphere, though primarily H_2O in composition, contains very little of the total water at the terrestrial surface, only about 0.00004 percent, yet the biosphere plays a major role in the transport of water vapour back into the atmosphere by the process of transpiration.

Earth's waters are not pure H_2O but contain dissolved and particulate materials. Thus, the masses of water at Earth's surface are major receptacles of inorganic and organic substances, and water movement plays a dominant role in the transportation of these substances about the planet's surface.

Biogeochemical Properties of the Hydrosphere

Rainwater

About 107,000 cubic km (nearly 25,800 cubic miles) of rain fall on land each year. The total water in the atmosphere is 13,000 cubic km, and this water, owing to precipitation and evaporation, turns over every 9.6 days. Rainwater is not pure but rather contains dissolved gases and salts, fine-ground particulate material, organic substances, and even bacteria. The sources of the materials in rainwater are the oceans, soils, fertilizers, air pollution, and fossil fuel combustion.

It has been observed that rains over oceanic islands and near coasts have ratios of major dissolved constituents very close to those found in seawater. The discovery of the high salt content of rain near coastlines was somewhat surprising because sea salts are not volatile, and it might be expected that the process of evaporation of water from the sea surface would "filter" out the salts. It has been demonstrated, however, that a large percentage of the salts in rain is derived from the bursting of small bubbles at the sea surface due to the impact of rain droplets or the breaking of waves, which results in the injection of sea aerosol into the atmosphere. This sea aerosol evaporates, with resultant precipitation of the salts as tiny particles that are subsequently carried high into the atmosphere by turbulent winds. These particles may then be transported over continents to fall in rain or as dry deposition.

Assuming equilibrium with the atmospheric carbon dioxide partial pressure (P_{CO_2}) of $10^{-3.5}$ (0.00035) atmosphere, the approximate mean composition of rainwater is in parts per million (ppm): sodium (Na^+), 1.98; potassium (K^+), 0.30; magnesium (Mg^{2+}), 0.27; calcium (Ca^{2+}), 0.09; chloride (Cl^-), 3.79; sulfate (SO_4^{2-}), 0.58; and bicarbonate (HCO_3^-), 0.12. In addition to these ions, rainwater contains small amounts of dissolved silica—about 0.30 ppm. The average pH value of rainwater is 5.6. (The term pH is defined as the negative logarithm of the hydrogen ion concentration in moles per litre. The pH scale ranges from 0 to 14, with lower numbers indicating increased acidity.) On a global basis, as much as 35 percent of the sodium, 55 percent of the chlorine, 15 percent of the potassium, and 37 percent of the sulfate in river water may be derived from the oceans through sea aerosol generation.

A considerable amount of data has become available for marine aerosols. These aerosols are important because (1) they are vital to any description of the global biogeochemical cycle of an element, (2) they may have an impact on climate, (3) they are a sink, via heterogeneous chemical reactions, for trace atmospheric gases, and (4) they influence precipitation of cloud and rain droplets. For many trace metals, the ratio of the atmospheric flux to the riverine flux for coastal and remote oceanic areas may be greater than one, indicating the importance of atmospheric transport. Figures have been prepared that illustrate the enrichment factors (EF) of North Atlantic marine aerosols and suspended matter in North Atlantic waters relative to the crust (that is, terrestrial sources), where,

$$EF_{crust} = \frac{(X/Al)_{air}}{(X/Al)_{crust}},$$

and $(X/Al)_{air}$ and $(X/Al)_{crust}$ refer, respectively, to the ratio of the concentration of the element X to that of Al, aluminum (which is an easily observed terrestrial component of aerosols), in the atmosphere and in average crustal material. Comparing the enrichment factors in marine aerosols with those of suspended matter in the water column indicates qualitatively the marine aerosols' importance as a source that alters the composition of marine suspended matter and, consequently, their importance to deep-sea sedimentation. Moreover, such comparisons help identify how significant terrestrial sources are for both the marine aerosols and the water below.

In some instances the ratios of ions in rainwater deviate significantly from those in seawater. Mechanisms proposed for this fractionation are, for example, the escape of chlorine as gaseous hydrogen chloride (HCl) from sea salt aerosol with a consequent enrichment in sodium and bubbling and thermal diffusion. In addition, release of biogenic gases such as dimethyl sulfide (DMS) from the sea surface and its subsequent reaction in the oceanic atmosphere to sulfate can change rainwater ion ratios with respect to seawater. Soil particles also can influence rainwater composition. Rainfall over the southwestern United States contains relatively high sulfate concentrations because of sulfate-bearing particles that have been blown into the atmosphere from desert soils. Rain near industrial areas commonly contains high contents of sulfate, nitrate, and carbon dioxide (CO_2) largely derived from the burning of coal and oil. There are two main processes leading to the conversion of sulfur dioxide (SO_2) to sulfuric acid (H_2SO_4). These are reactions with hydroxyl radicals ($-OH$) and with hydrogen peroxide (H_2O_2) in the atmosphere:

$$SO_2 + CH \rightarrow intermediate\ species \rightarrow H_2SO_4$$

and,

$$SO_2 + H_2O_2 = H_2SO_4 .$$

The sulfuric acid then dissociates to hydrogen and sulfate ions:

$$H_2SO_4 = 2H^+ + SO_4^{2-}$$

For the nitrogen gases nitric oxide (NO) and nitrogen dioxide (NO_2) released from fossil fuel burning, their atmospheric reactions lead to the production of nitric acid (HNO_3) and its dissociation to hydrogen ions (H^+) and nitrate (NO_3^-). These reactions are responsible for the acid rain conditions that occurred in the northeastern United States, southeastern Canada, and western Europe during the second half of the 20th century. The high sulfate values of the rain in the northeastern United States reflect the acid precipitation conditions of this region.

River and Ocean Waters

River discharge constitutes the main source for the oceans. Seawater has a more uniform composition than river water. It contains, by weight, about 3.5 percent dissolved salts, whereas river water has only 0.012 percent. The average density of the world's oceans is roughly 2.75 percent greater than that of typical river water. Of the average 35 parts per thousand salts of seawater, sodium and chlorine make up almost 30 parts, and magnesium and sulfate contribute another four parts. Of the remaining one part of the salinity, calcium and potassium constitute 0.4 part each and carbon, as carbonate and bicarbonate, about 0.15 part. Thus, nine elements (hydrogen, oxygen, sulfur, chlorine, sodium, magnesium, calcium, potassium, and carbon) make up 99 percent of seawater, though most of the 94 naturally occurring elements have been detected therein. Of importance are the nutrient elements phosphorus, nitrogen, and silicon, along with such essential micronutrient trace elements as iron, cobalt, and copper. These elements strongly regulate the organic production of the world's oceans.

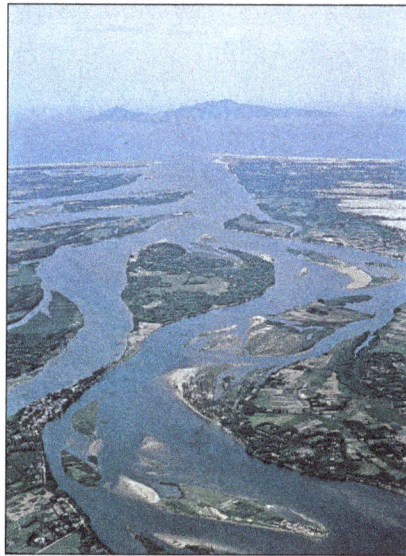

A portion of the delta of the Mekong River as it flows through southern Vietnam and empties into the South China Sea.

In contrast to ocean water, the average salinity of the world's rivers is low—only about 0.012 percent, or 120 ppm by weight. Of this salt content, carbon as bicarbonate constitutes 58 parts, or 48 percent, and calcium, sulfur as sulfate, and silicon as dissolved monomeric silicic acid make up a total of about 39 parts, or 33 percent. The remaining 19 percent consists predominantly of chlorine, sodium, and magnesium in descending importance. It is obvious that the concentrations and relative proportions of dissolved species in river waters contrast sharply with those of seawater. Thus, even though seawater is derived in part by the chemical differentiation and evaporation of river water, the processes involved affect every element differently, indicating that simple evaporation and concentration are entirely secondary to other processes.

Water-rock Interactions as Determining River Water Composition

Generally speaking, the composition of river water, and thus that of lakes, is controlled by water-rock interactions. The attack of carbon dioxide-charged rain and soil waters on the individual minerals in continental rocks leads to the production of dissolved constituents for lakes, rivers, and streams. It also gives rise to solid alteration products that make up soils or suspended particles in freshwater aquatic systems. The carbon dioxide content of rain and soil waters is of particular importance in weathering processes. The pH of rainwater equilibrated with the atmospheric carbon dioxide partial pressure of $10^{-3.5}$ (0.00032) atmosphere is 5.6. In industrial regions, rainwater pH values may be lower because of the release and subsequent hydrolysis of acid gases—namely, sulfur dioxide and nitrogen oxides (NO_x) from the combustion of fossil fuels. After rainwater enters soils, its characteristics change markedly. The usual few parts per million of salts in rainwater increase substantially as the water reacts. The upper part of the soil is a zone of intense biochemical activity. The bacterial population near the surface is large, but it decreases rapidly downward. One of the major biochemical processes of the bacteria is the oxidation of organic material, which leads to the release of carbon dioxide. Soil gases obtained above the zone of water saturation may contain 10 to 40 times as much carbon dioxide as the free atmosphere, and in some cases carbon dioxide has been shown to make up 30 percent of the soil gases as opposed to 0.03 percent of the free atmosphere. In addition to the acid effects of carbon dioxide, there is a highly acidic microenvironment created by the roots of living plants. Values of pH as low as 2 have been measured immediately adjacent to root hairs. The combined length of a plant's root hairs may be several kilometres, so their chemical effects on acidic water are formidable.

Suspended soil particles. The brown colour of the water is the result of rains washing soil particles into the stream from surrounding land and the stirring up of mud on the riverbed.

Congruent and Incongruent Weathering Reactions

These acid solutions in the soil environment attack the rock minerals, the bases of the system, producing neutralization products of dissolved constituents and solid particles. Two general types of reactions occur: congruent and incongruent. In the former, a solid dissolves, adding elements to the water according to their proportions in the mineral. An example of such a weathering reaction is the solution of calcite ($CaCO_3$) in limestones:

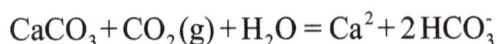

$$CaCO_3 + CO_2(g) + H_2O = Ca^2 + 2HCO_3^-$$

Here one of the HCO3– ions comes from calcite and the other from CO2(g) in the reacting water. The amount of carbon dioxide dissolved according to reaction $CaCO_3 + CO_2(g) + H_2O = Ca^2 + 2HCO_3^-$

depends on temperature, pressure, original bicarbonate content of the weathering solution, and partial pressure of the carbon dioxide. The carbon dioxide and the temperature are the most important variables. Increases in one or both of these variables lead to increases in the amount of calcite dissolved. For example, for a carbon dioxide pressure of 10–3.5 (0.00032) atmosphere, the amount of calcium that can be dissolved until saturation is about 10–3.3 (0.0005) mole, or 20 ppm, at 25 °C (77 °F). For an atmospheric carbon dioxide pressure of 10–2 (0.01) atmosphere and for a soil atmosphere of nearly pure carbon dioxide, the values are 65 and 300 ppm, respectively. The weathering of calcite leads to the release of calcium and bicarbonate ions into soil waters and groundwater, and these constituents eventually reach lake and river systems. The insoluble residue of quartz (SiO_2), clay minerals, and iron oxides (e.g., FeOOH) in the limestone rock make up the deep-red soils that form from limestone weathering. These particles may be carried into streams by runoff and hence to lakes and the oceans and become part of the suspended load of these systems.

An example of an incongruent weathering reaction—that is, one where only part of a solid is consumed—is that involving aluminosilicates. One such reaction is the aggressive attack of carbon dioxide-charged soil water on the mineral K-spar ($KAlSi_3O_8$), an important phase found in continental rocks. The reaction is:

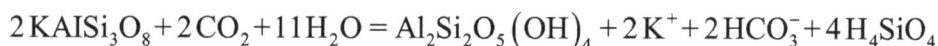

$$2\,KAlSi_3O_8 + 2\,CO_2 + 11\,H_2O = Al_2Si_2O_5(OH)_4 + 2\,K^+ + 2\,HCO_3^- + 4\,H_4SiO_4$$

It should be noted that the K-spar changes into a new mineral—kaolinite (a clay mineral) in this case—plus solution, and acid is consumed. The total dissolved material per litre of soil solution released is about 60 ppm for a solution initially equilibrated with a typical soil carbon dioxide content. The water resulting from reaction $2\,KAlSi_3O_8 + 2\,CO_2 + 11\,H_2O = Al_2Si_2O_5(OH)_4 + 2\,K^+ + 2\,HCO_3^- + 4\,H_4SiO_4$ would contain bicarbonate, potassium, and dissolved silica in the ratios 1:1:2, and the new solid, kaolinite, would be a weathering product. These dissolved constituents and the solid alteration product would eventually reach rivers to be transferred possibly to lakes and ultimately to the sea. It has been demonstrated that the composition of river water is the product of a variety of mineral-water reactions such as $CaCO_3 + CO_2(g) + H_2O = Ca^2 + 2\,HCO_3^-$ and $2\,KAlSi_3O_8 + 2\,CO_2 + 11\,H_2O = Al_2Si_2O_5(OH)_4 + 2\,K^+ + 2\,HCO_3^- + 4\,H_4SiO_4$. The dissolved load of the world's rivers comes from the following sources: 7 percent from beds of halite (NaCl) and salt disseminated in rocks, 10 percent from gypsum ($CaSO_4 \cdot 2H_2O$) and anhydrite ($CaSO_4$) deposits and sulfate salts disseminated in rocks, 38 percent from limestones and dolomites, and 45 percent from the weathering of one silicate mineral to another. Of the bicarbonate ions in river water, 56 percent stems from the atmosphere, 35 percent from carbonate minerals, and 9 percent from the oxidative weathering of fossil organic matter. Reactions involving silicate minerals account for 30 percent of the riverine bicarbonate ions.

Besides dissolved substances, rivers also transport solids in traction (i.e., bed load) and, most importantly, suspended load. The present global river-borne flux of solids to the oceans is estimated as 15.5 billion metric tons (about 17.1 billion tons) per year. Present elemental fluxes are estimated in millions of metric tons per year as silicon, 4,420; aluminum, 1,460; iron, 740; calcium, 330; potassium, 310; magnesium, 210; and sodium, 110. The total load of particulate organic carbon of the world's rivers is 180 million metric tons (nearly 200 million tons) per year. The riverine fluxes of trace metals to the oceans are dominated by their occurrence in the particulate phase as opposed to the dissolved phase. The particulate matter in river water is an important source of silicon, aluminum, iron, titanium, rubidium, scandium, vanadium, the lanthanoids, and other elements for deep-sea sediments.

Lake Waters

Although lake waters constitute only a small percentage of the water in the hydrosphere, they are an important ephemeral storage reservoir for fresh water. Aside from their recreational use, lakes constitute a source of water for household, agricultural, and industrial uses. Lake waters are also very susceptible to changes in chemical composition due to these uses and to other factors.

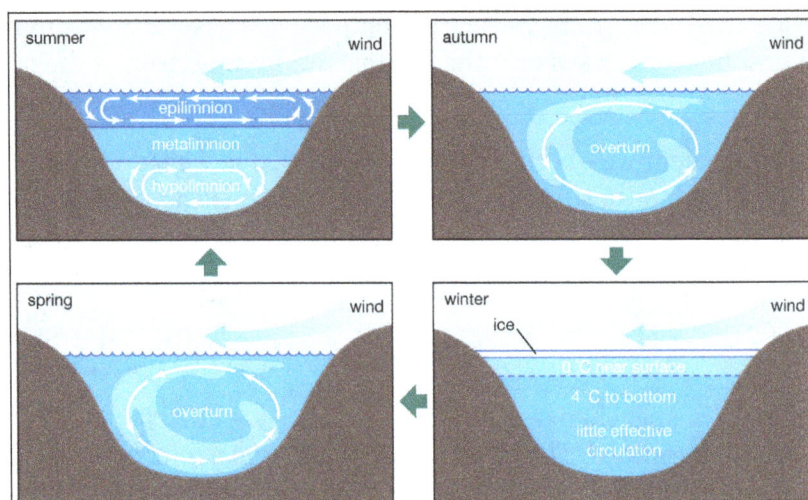

Temperate lake circulation.

Annual circulation patterns in a dimictic lake. The typical dimictic lake has distinct layers that fully mix twice a year. It undergoes stratification in the summer and complete overturn in the autumn and spring. During winter, surface ice prevents further mixing by the wind. Small differences in density and temperature exist, with cooler water (0 °C [32 °F]) staying near the surface and warmer, denser water (4 °C [39.2 °F]) extending to the bottom.

In general, fresh waters at the continental surface evolve from their rock sources by enrichment in calcium and sodium and by depletion in magnesium and potassium. In very soft waters the alkalies may be more abundant than the alkaline earths, and in the more-concentrated waters of open river systems calcium > magnesium > sodium > potassium. For the anions, in general, HCO_3^- exceeds SO_4^{2-}, which is greater in concentration than Cl^-. It is worthwhile at this stage to consider some major mechanisms that control global surface water composition. These mechanisms are atmospheric precipitation, rock reactions, and evaporation-precipitation.

The mechanism principally responsible for waters of very low salinity is precipitation. These waters tend to form in tropical regions of low relief and thoroughly leached source rocks. In these regions rainfall is high, and volumes of fresh water (rivers, tributary streams, pools, etc.) within a watershed are usually dominated by salts brought in by precipitation. Such waters constitute one part of a chain of water volumes that begins with falling precipitation and ends with the release of water into the ocean, for which the final part of the chain represents water volumes dominated by contributions of dissolved salts from the rocks and soils of their basins. These waters have moderate salinity and are rich in dissolved calcium and bicarbonate. They are in turn the "end-member" of another series that extends from the calcium-rich, medium-salinity fresh waters to the high-salinity, sodium chloride-dominated waters of which seawater is an example. Seawater composition, however, does not evolve directly from the composition of fresh waters and the precipitation of calcium carbonate; other mechanisms that control its composition are involved. Such factors as relief and vegetation also may affect the composition of the world's surface waters, but atmospheric precipitation, water-rock reactions, and evaporation-crystallization processes appear to be the dominant mechanisms governing continental surface water chemistry.

Continental fresh waters evaporate once they have entered closed basins, and their constituent salts precipitate on the basin floors. The composition of these waters may evolve along several different paths, depending on their initial chemical makeup.

Biological processes strongly affect the composition of lake waters and are responsible to a significant degree for the compositional differences between the upper water layer (the epilimnion) and the lower water layer (the hypolimnion) of lakes. The starting point is photosynthesis, represented by the following reaction:

$$106\,CO_3 + 16\,NO_3^- + HPO_4^{2-} + 122\,H_2O + 18\,H^+ \xrightarrow[\substack{\text{micro-}\\\text{nutrients}}]{\substack{\text{radiant}\\\text{energy}}} C_{106}H_{263}O_{110}N_{16}P + 138\,O_2$$

The reversal of this reaction is oxidation-respiration leading to the release of the nutrients nitrogen and phosphorus, as well as carbon dioxide. In a stratified lake, carbon, nutrients, and silica are extracted from the upper layer during photosynthesis. This process leads to reduced concentrations of nitrate, phosphate, and silica in these waters and, during times of maximum daylight organic production, to supersaturation of the upper layer with respect to dissolved oxygen. The organic matter produced by phytoplankton may be either grazed upon by zooplankton and other organisms or decomposed by bacteria. Some of it, however, sinks into the lower layer. There it is further decomposed, especially by bacteria, resulting in the release of dissolved phosphorus and nitrogen and the consumption of oxygen. Oxygen concentrations therefore are reduced in these lower lake waters, because stratification prevents oxygen exchange with the atmosphere. Furthermore, the inorganic carbonate and siliceous skeletons of the dead organisms sinking into the lower layer may dissolve, giving rise to increased concentrations of dissolved silica and inorganic carbon in the deep waters of stratified lakes. This dissolution is a result of undersaturation of the waters of the lower layer with respect to the opaline silica and calcium carbonate that make up the skeletons of the dead and sinking plankton. These natural biological processes have been accelerated in some lakes because of excess nutrient input by human activity, resulting in the eutrophication of lake waters and marine systems.

Groundwaters

Groundwaters derive their compositions from a variety of processes, including dissolution, hydrolysis, and precipitation reactions; adsorption and ion exchange; oxidation and reduction; gas exchange between groundwater and the atmosphere; and biological processes.

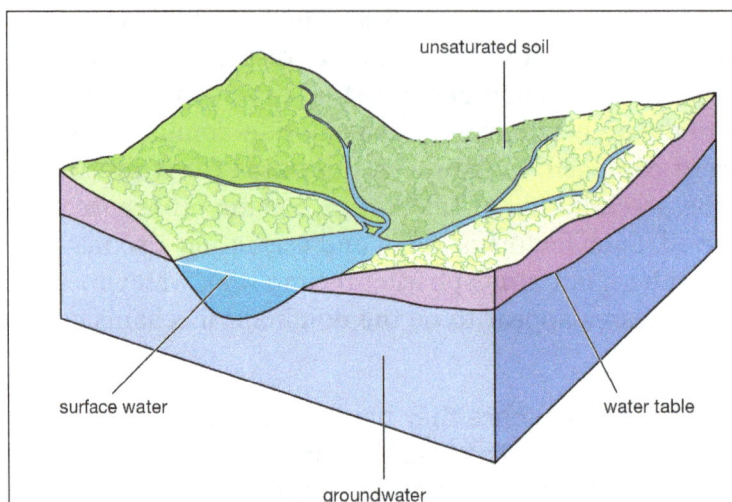

The water table is the top level of groundwater.
Surface water is an exposed part of the water table.

The biological processes of greatest importance are microbial metabolism, organic production, and respiration (oxidation). By far the most important overall process for the major constituents of groundwater is that of mineral-water reactions, which were briefly described above in River and ocean waters. Thus, the composition of groundwaters strongly reflects the types of rock minerals that the waters have encountered in their movement through the subsurface.

In general, the most mobile elements in groundwater—i.e., those most easily liberated by the weathering of rock minerals—are calcium, sodium, and magnesium. Silicon and potassium have intermediate mobilities, and aluminum and iron are essentially immobile and locked up in solid phases.

Groundwaters are highly susceptible to contamination because of human activities and the fact that their dissolved constituents are derived to a large extent from the leaching of surface materials. Some of the nitrogen and phosphorus applied to soils as fertilizers and organic pesticides may be leached and leak into groundwater systems, leading to increased concentrations of ammonium and phosphate. Radioactive wastes, industrial chemicals, household materials, and mine refuse are other anthropogenic sources of dissolved substances that have been detected in groundwater systems.

Ice

Ice is nearly a pure solid and, as such, accommodates few foreign ions in its structure. It does contain, however, particulate matter and gases, which are trapped in bubbles within the ice. The change in composition of these materials through time, as recorded in the successive layers of ice, has been used to interpret the history of Earth's surface environment and the impact of human activities on this environment. The increase in the lead content of continental glacial ice with decreasing age of the ice up to the middle of the 1970s, for example, reflects the progressive input of tetraethyl lead into the global environment from gasoline burning. (Stringent environmental regulations that appeared in the 1970s regarding the use of leaded gasoline has led to a fall in lead concentrations in ice laid down since that time.) Also, atmospheric carbon dioxide and methane concentrations, which have increased significantly during the past century because of anthropogenic activities, are faithfully recorded in ice bubbles of the thick continental ice sheets.

By 2016 atmospheric carbon dioxide and methane concentrations had increased by more than 43 percent and more than 150 percent, respectively, higher than their concentrations 200 years ago; the latter concentration values were obtained from measurements of the gases in air trapped in ice.

Los Glaciares National Park. Although many other South American glaciers have declined in mass since the middle of the 20th century, the mass of Perito Moreno Glacier has remained relatively steady.

The Water Cycle

General Nature of the Cycle

The present-day water cycle at Earth's surface is made up of several parts. Some 496,000 cubic km (about 119,000 cubic miles) of water evaporates from the land and ocean surface annually, remaining for about 10 days in the atmosphere before falling as rain or snow. The amount of solar radiation necessary to evaporate this water is half of the total solar radiation received at Earth's surface. About one-third of the precipitation falling on land runs off to the oceans primarily in rivers, while direct groundwater discharge to the oceans accounts for only about 0.6 percent of the total discharge. A small amount of precipitation is temporarily stored in the waters of rivers and lakes. The remaining precipitation over land, 73,000 cubic km (17,500 cubic miles) per year, returns to the atmosphere by evaporation. Over the oceans, evaporation exceeds precipitation, and the net difference represents transport of water vapour over land, where it precipitates as rain or snow and returns to the oceans as river runoff and direct groundwater discharge.

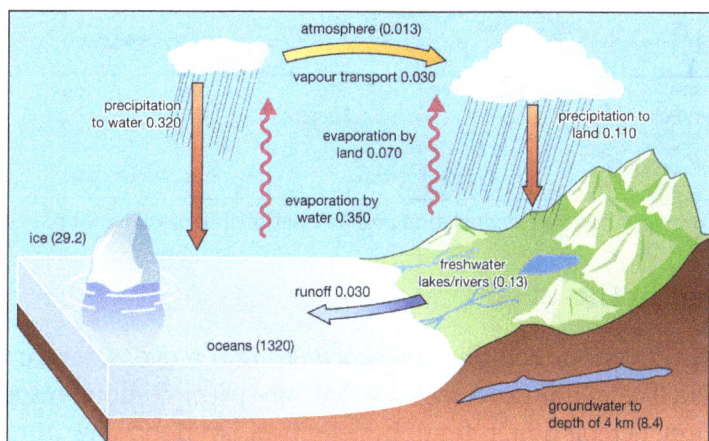

The present-day surface hydrologic cycle, in which water is transferred from the oceans through the atmosphere to the continents and back to the oceans over and beneath the land surface. The values in parentheses following the various forms of water (e.g., ice) refer to volumes in millions of cubic kilometres; those following the processes (e.g., precipitation) refer to their fluxes in millions of cubic kilometres of water per year.

The various reservoirs in the water cycle have different water residence times. Residence time is defined as the amount of water in a reservoir divided by either the rate of addition of water to the reservoir or the rate of loss from it. The oceans have a water residence time of 3,000 to 3,230 years; this long residence time reflects the large amount of water in the oceans. In the atmosphere the residence time of water vapour relative to total evaporation is only about 10 days. Lakes, rivers, ice, and groundwaters have residence times lying between these two extremes and are highly variable.

There is considerable variation in evaporation and precipitation over the globe. In order for precipitation to occur, there must be sufficient atmospheric water vapour and enough rising air to carry the vapour to an altitude where it can condense and precipitate. Precipitation and evaporation vary with latitude and their relation to the global wind belts. The trade winds, for example, are initially cool, but they warm up as they blow toward the Equator. These winds pick up moisture from the ocean, increasing ocean surface salinity and causing seawater at the surface to sink. When the trade winds reach the Equator, they rise, and the water vapour in them condenses and forms clouds. Net precipitation is high near the Equator and also in the belts of the prevailing westerlies, where there is frequent storm activity. Evaporation exceeds precipitation in the subtropics, where the air is stable, and near the poles, where the air is both stable and has a low water vapour content because of the cold. The Greenland Ice Sheet and the Antarctic Ice Sheet formed because the very low evaporation rates at the poles resulted in precipitation exceeding evaporation in these local regions.

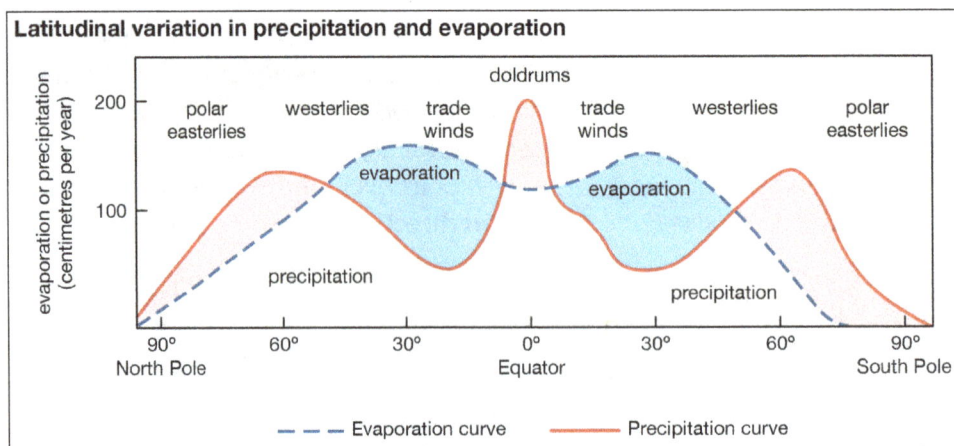

Latitudinal variation in precipitation and evaporation and its relationship to major wind belts and oceanic salinity.

Processes Involved in the Cycle

The water cycle consists of various complicated processes that move water throughout the different reservoirs on the planet. The major processes involved are precipitation, evaporation, interception, transpiration, infiltration, percolation, retention, detention, overland flow, throughflow, and runoff.

Water Vapour and Precipitation

Water exists in the atmosphere in gaseous form. Its liquid form, either as water droplets in clouds or as rain, and its solid form, as ice crystals in clouds, snowflakes, or hail, occur only momentarily and locally.

Water vapour performs two major functions: (1) it is important to the radiation balance of Earth, as its presence keeps the planetary surface warmer than would otherwise be the case, and (2) it is the principal phase of the ascending part of the water cycle.

The mass of water vapour in the atmosphere, which represents only 0.001 percent of the hydrosphere, is highest in the tropics and decreases toward the poles. At a mean temperature of Earth's surface of 15 °C (59 °F), the partial pressure of water vapour at equilibrium with pure water is 0.017 atmosphere. The addition of salts to pure water lowers its vapour pressure. The equilibrium, or saturation, water vapour pressure of a saturated solution of sodium chloride is 22 percent lower than that of pure water. Precipitable water vapour has, on the average, a vapour pressure of 0.0025 atmosphere, which amounts to 15 percent of the saturation vapour pressure. The ratio of observed water vapour pressure to the saturation vapour pressure at the same temperature is the relative humidity of the air. Thus, the mean relative humidity of the atmosphere is only 15 percent, a value that is low in human terms, in which levels of 50 to 60 percent are preferred for maximum comfort. The relative humidity of the air, however, varies greatly from one geographic region to another and also vertically in the atmosphere. Atmospheric water vapour decreases rapidly with increasing altitude relative to its surface value. The amount of water required to saturate a volume of air depends on the temperature of the air. Air at high temperature can hold more water vapour at saturation than can air at low temperature. Because the temperature of the lower atmosphere (the troposphere) decreases rapidly with increasing altitude to about 15 km (about 9 miles), the upper levels of the troposphere contain little water vapour; most of the vapour is found within a few kilometres of Earth's surface. The average relative humidity of tropospheric air is about 50 percent. Above 15 km, water vapour is essentially frozen out of the atmosphere, amounting to less than 0.1 percent of its concentration at Earth's surface.

Aside from temperature, other factors determine the water vapour content of the air and are particularly important in the lower troposphere. These factors include local evaporation and the horizontal atmospheric transportation of moisture, which varies with altitude, latitude, season, and topography. During a period of 10 days (i.e., the residence time of water in the atmosphere), horizontal eddy turbulence may disperse vapour over distances up to 1,000 km (600 miles).

When a mass of air at Earth's surface is exposed to a body of water, it gains water by evaporation or loses water by precipitation, depending on its relative humidity. If the air is undersaturated, with a relative humidity of less than 100 percent, it gains water vapour because the rate of evaporation exceeds the rate of condensation. If the air is supersaturated, with a relative humidity greater than 100 percent, the air mass loses water vapour because the rate of precipitation exceeds that of evaporation. This interaction between air masses and surface water bodies drives the atmosphere toward a state of saturation, which is not achieved for the entire atmosphere because of the variability in weather and because not all air masses are in contact with water bodies. In general, the level of atmospheric water vapour is higher in the summer, since temperatures are higher at this time of year. Also, atmospheric water vapour content is higher near the source of moisture than in distant regions. Over the oceans, the air is almost always near saturation, whereas over the deserts, where the supply of moisture is limited, the air is far below water vapour saturation values. In most cases, atmospheric water vapour content decreases inland over continents, but this decrease is modified by rainfall conditions, by the presence or absence of high mountains, large lakes, extensive forests, and swamps, and by the prevailing wind directions. Horizontal winds and air mass movements transfer water vapour from the ocean to the land. Although the processes are not completely separable, the horizontal transfer of water vapour seldom causes the vapour to undergo condensation, whereas vertical movements are most important in the condensation process.

Condensation depends strongly on the average temperature of Earth's surface because the water vapour content of the air is strongly dependent on temperature. In figures that show the states of water as a function of the variables of pressure and temperature, the slope of the phase boundary between liquid water and water vapour is positive, implying that with increasing temperature the air at equilibrium will hold increasing amounts of water vapour. Cooling or mixing of this air results in condensation of the vapour and precipitation as water droplets or as ice crystals if the air temperature is below 0 °C (32 °F). When first formed, the water droplets or ice crystals are very small, on the order of 10^{-2} to 10^{-3} cm (0.004 to 0.0004 inch) in diameter, and they float freely in the atmosphere. In large quantities, these water droplets and ice crystals produce a cloud. All clouds are formed as a result of cooling below the dew point, the temperature at which condensation begins when air is cooled at constant pressure and constant water vapour content. When the droplets or crystals coalesce to a size of about 10^{-2} cm (0.004 inch) in diameter, they become heavy enough to fall as raindrops or snowflakes. Hailstones measure about 10^{-1} cm (0.04 inch) in diameter or much larger. Water vapour condensing in the atmosphere contains strongly soluble salts (mostly of oceanic origin), weakly soluble or insoluble solids (dust), and dissolved gases. The dust and sea salt aerosol particles in the air may act as sites of condensation by serving as nuclei for bringing initially a few water molecules together and inducing condensation from supersaturated air.

Distribution of Precipitation

Precipitation falling toward Earth's surface may suffer several fates. It may be evaporated during its fall or after it reaches the ground surface. If the surface is covered with dense vegetation, much of the precipitation may be held on leaves and plant limbs and stems. This process is termed interception and may result in little water reaching the ground because the water may be directly evaporated from plant surfaces back into the atmosphere. If precipitation reaches the ground in the form of snow, it may remain there for some time. On the other hand, if precipitation falls as rain, it may evaporate, infiltrate the soil, be detained in small catchment areas, or become overland flow—a form of runoff. Overland flow (R_o) may be expressed in terms of intensity units, water depth per unit of time (e.g., centimetres per hour, or inches per hour), as:

$$R_o = P - I,$$

where P is precipitation rate and I is infiltration rate (rate of entry and downward movement of water into the soil profile). Infiltration rate will equal precipitation rate until the limit of the infiltration rate, or infiltration capacity, is reached. Soil infiltration rates are usually high at the beginning of a rain preceded by a dry spell and decrease as the rainfall continues. This change in rate is due to the clogging of soil pores by particles brought from above by the infiltrating rain and to the swelling of colloidal soil particles as they absorb water. Thus, rapid decreases in infiltration rates during a rain are more likely to occur in clay-rich soils than in sandy soils.

Between rainfall periods, water held in the soil as moisture is gradually lost by direct evaporation or by withdrawal by plants. Evaporation into the open atmosphere occurs at the surface of the soil, and the soil dries progressively downward with time. Water vapour in the soil diffuses upward, replenishing the evaporated water, and in turn is evaporated. The pumping of air into and out of the soil by atmospheric pressure changes enhances the movement of soil moisture upward. It has been shown that evaporation of a water droplet in the free atmosphere, and to a first approximation in various soil atmospheres, is proportional to the droplet surface area $4\pi r^2$ (square centimetres, where r is the radius of the droplet), the diffusional flux of water at the droplet surface, and the transfer of heat as the droplet evaporates. The equation for the rate of shrinkage of a water droplet due to evaporation is:

$$\frac{dr}{dt} = -\frac{D\rho_{vo}(Sp-1)}{r\rho_L(1+X)},$$

where dr/dt is the rate of change in the radius of the water droplet (centimetres per second), D is the diffusion coefficient of water vapour in air (cubic centimetres per second), ρ_{vo} is the equilibrium vapour concentration at the droplet surface, Sp is the degree of undersaturation of water vapour in the environment, ρ_L is the density of liquid water (grams per cubic centimetre), and X is a dimensionless parameter depending on D, ρ_{vo}, temperature, the heat of evaporation of water vapour, the coefficient of thermal conductivity of air, and the spherical coordinate system necessary to define processes occurring to a spherical water droplet. Water droplets shrink—dr/dt < 0, evaporate—when the water vapour concentration in the environment (atmosphere or soil atmosphere) is less than the saturation water vapour concentration at the droplet surface. They grow—dr/dt > 0, condense—when the converse is true in the free atmosphere. The term dr/dt has negative values for evaporation and positive ones for condensation. Use of this equation shows, as an example, that it would take 23 minutes for a water droplet to shrink (evaporate) in size from 50 to 5 micrometres (0.002 to 0.0002 inch) in air at 10 °C (50 °F) and a water vapour undersaturation of 1 percent.

Besides simple evaporation of water from soils, water is also returned to the atmosphere by transpiration in plants. Plants draw water from soil moisture through their vast network of root hairs and rootlets. This water is carried upward through the plant trunk and branches into the leaves, where it is discharged as water vapour. The term evapotranspiration is used in climatic and hydrologic studies to include the combined water loss from Earth's surface resulting from evaporation and transpiration. The maximum possible evapotranspiration is termed potential evapotranspiration and is governed by the available heat energy. It is taken as equal to evaporation from a large water surface and is generally much less than actual evapotranspiration. Actual evapotranspiration is never greater than precipitation except on irrigated land because of percolation of water into groundwater bodies and surface runoff.

The soil moisture zone gains water by precipitation and infiltration and loses water by evapotranspiration, overland flow, and percolation of water downward due to gravity into the groundwater zone. The contact between the groundwater zone (phreatic zone) and the overlying unsaturated zone (vadose zone) is called the groundwater table. The water balance equation for change of moisture storage in a soil is given as:

$$S = P - E - R,$$

where S is storage, P is precipitation, E is evaporation, and R is surface runoff plus percolation rate into the groundwater zone; all terms are in units of length per unit of time (e.g., millimetres per day, centimetres per month). In humid midlatitude climates where a strong contrast between winter and summer temperatures exists, there is an annual cycle of the water content of soils. The annual cycle of moisture in soil in Ohio, U.S., for example, demonstrates the processes controlling soil moisture. Of special importance is the fact that the soils are saturated in this temperate climate in the spring, and the evaporation rate is low because of the low input of radiant energy from the Sun. By contrast, in the summer, evaporation increases because of increasing solar radiation, and with the growth of plants so does transpiration. Soil moisture is reduced to very low levels at this time of year.

Groundwaters and River Runoff

The term R in the water balance equation for change of soil moisture storage above represents groundwater and river runoff losses from the soil moisture zone. Water percolates from the soil moisture zone through the unsaturated (vadose) zone to the water table. Flow through the unsaturated zone is complicated. After a rainfall, water may form a nearly continuous phase in pores in this zone, but, with drying, the last amount of water is held in clusters at points of contact of solid grains and as thin films on solid surfaces. The flow paths of water become more tortuous, and the water-conducting properties decrease rapidly. Structured soils and fractured rock in the vadose zone may act as conduits for fluids to reach the water table. Because of the complex geometry of water contained in the unsaturated zone, the properties of water are expressed by means of empirical relationships. Darcy's law, derived in 1856 from experimentation by the French engineer Henri Darcy, permits quantification of water flow through porous media. The law states that the rate of flow Q of a fluid through a porous layer of medium (e.g., a sand bed) is directly proportional to the area A of the layer and to the difference Δh between the fluid heads at the inlet and outlet faces of the layer and is inversely proportional to the thickness L of the layer. Expressed analytically,

$$Q = \frac{KA\Delta h}{L},$$

where, K is a constant characteristic of the medium. The term K for a porous rock medium is the volume of fluid of unit viscosity passing through a unit cross section of the rock in unit time under the action of a unit pressure gradient; this characteristic is called permeability. The permeability of a rock is dependent on the geometric properties of the rock, such as porosity, shape and size distribution of constituent rock grains, and degree of cementation of the rock. Permeabilities of rocks vary greatly. Unconsolidated sands may have permeabilities measured in hundreds of darcys, whereas consolidated sands that will transmit reasonable amounts of fluid have permeabilities of 0.01 to 1 darcy. A rough idea of the meaning of one darcy of permeability (which equals 9.869×10^{-12} square metre [1.261×10^{-11} square foot]) can be obtained by imagining a cube of sand one foot on a side. If the sand has a permeability of one darcy, approximately one barrel of water per day will pass through the one-foot cube with a one-pound pressure head. The general equation of Darcy can be modified to express flow in both the unsaturated zone and the saturated groundwater zone.

Groundwater is constantly in motion. When a lake or stream intersects the groundwater table, groundwater communicates directly with these bodies of water. If the groundwater table is higher than the stream or lake level, a pressure head will develop such that the groundwater flows into the water body; conversely, if the groundwater table is lower than the river or lake level, the pressure gradient induces flow into the groundwater. Most groundwater ultimately reaches the channels of surface streams and rivers and flows to the sea. On the average, groundwater contributes to total river runoff about 30 percent of its water on a global basis.

Water runoff from the land surface is that part of precipitation which eventually appears in perennial or intermittent surface streams. Streamflow-generation mechanisms have been studied for several decades, and there is now considerable knowledge regarding rainfall runoff processes and their controls. This understanding is the result of both careful observations from field experiments and the heuristic simulations of hypothetical realities with rigorous mathematical models. The discharge measured at the downstream end of a channel reach is supplied by channel inflow at the upstream end of the reach and by the lateral inflows that enter the channel from the hillslope along the reach. The lateral inflows may arrive at the stream in one of three forms: (1) groundwater flow, (2) subsurface storm flow, or (3) overland flow.

Groundwater flow provides the base flow component of streams that sustains their flow between storms. The "flashy" response of streamflow to individual precipitation events may be ascribed to either subsurface storm flow or overland flow. Under intense rainfall events during which the surface soil layer becomes saturated to some depth, water is able to migrate through "preferred pathways" rapidly enough to deliver contributions to the stream during the peak runoff period. The conditions for subsurface storm flow are quite restrictive. The mechanism is most likely to be operative on steep, humid, forested hillslopes with very permeable surface soils.

Overland flow is generated at a point on a hillslope only after surface ponding takes place. Ponding cannot occur until the surface soil layers become saturated. It is now widely recognized that surface saturation can occur because of two quite distinct mechanisms—specifically, Horton overland flow (named for American hydraulic engineer and hydrologist Robert E. Horton) and Dunne overland flow (named for British hydrologist Thomas Dunne).

The former classic mechanism is for a precipitation rate that exceeds the saturated hydraulic conductivity of the surface soil. A moisture content versus depth profile during such a rainfall event will show moisture contents that increase at the surface as a function of time. At some point in time the surface becomes saturated, and an inverted zone of saturation begins to propagate downward into the soil. It is at this time that the infiltration rate drops below the rainfall rate and overland flow is generated. The time is called the ponding time. The necessary conditions for the generation of overland flow by the Horton mechanism are (1) a rainfall rate greater than the saturated hydraulic conductivity of the soil and (2) a rainfall duration longer than the required ponding time for a given initial moisture profile. Horton overland flow is generated from partial areas of the hillslope where surface hydraulic conductivities are lowest.

In Dunne overland flow, the precipitation rate is less than the saturated hydraulic conductivity, and the initial water table is shallow or there is a shallow impeding layer. Surface saturation occurs because of a rising water table; ponding and overland flow occur at a time when no further soil moisture storage is available. The Dunne mechanism is more common to near-channel areas. Dunne overland flow is generated from partial areas of the hill slope where water tables are shallowest. Both Horton and Dunne mechanisms result in variable source areas that expand and contract through wet and dry periods.

Total river discharge and the chemistry of the discharge vary from continent to continent; some continents are wetter and some drier than the world average, but the deviations are not extreme. The runoff per unit area from Asia and Europe is almost exactly equal to the world average; it is a little lower in Africa and North America; and it is considerably higher in South America. Antarctica is frozen and Australia is arid, and so they contribute little runoff. Also, since their areas are relatively small, they do not affect the global runoff average significantly. The waters draining the continents have quite different chemistries; those from Europe are very rich in calcium and bicarbonates, whereas those from Africa and South America are not. North American and Asian rivers are somewhat intermediate in their concentrations of these dissolved constituents. Such differences in composition reflect a variety of factors, including runoff, temperature, and relief, but certainly the bulk composition of the continental rocks in contact with these waters and their underground sources play a major role. The surface rocks of Europe are rich in carbonates, and those of South America are not; the latter are dominated by sediments rich in silicate minerals.

The chemistry of groundwater and river runoff is being modified by human activities on a global scale. The natural dissolved riverine input of major constituents to the oceans already has been increased by more than 10 percent because of human activities. In the case of sodium, chlorine, and sulfate, the increases are as high as 30 percent. In the United States alone, total water use is equivalent to one-third of total runoff, with about 2 percent of the water used coming from underground wells. In the southwestern region of the country, water supplies have been tapped heavily and in some areas have been exhausted with no hope of replacement. This extensive use of fresh waters in the United States and throughout the globe makes them particularly susceptible to pollution. Leachates from fertilizers, herbicides, and pesticides are found in some freshwater bodies; toxins or excessive amounts of certain inorganic or organic chemicals are present; radioactive elements have been detected; and some surface water bodies have had their salinities increased dramatically, rendering them useless for human consumption. It is therefore imperative that countries closely monitor the use of freshwater systems and promote their conservation.

Origin and Evolution of the Hydrosphere

It is not very likely that the total amount of water at Earth's surface has changed significantly over geologic time. Based on the ages of meteorites, Earth is thought to be 4.6 billion years old. The oldest rocks known are 3.9 billion to 4.0 billion years old, and these rocks, though altered by post-depositional processes, show signs of having been deposited in an environment containing water. There is no direct evidence for water for the period between 4.6 billion and 3.9–4.0 billion years ago. Thus, ideas concerning the early history of the hydrosphere are closely linked to theories about the origin of Earth.

Earth is thought to have accreted from a cloud of particles around the Sun. This gaseous matter condensed into small particles that coalesced to form a protoplanet, which in turn grew by the gravitational attraction of more particulates. Some of these particles had compositions similar to that of carbonaceous chondrite meteorites, which may contain up to 20 percent water. Heating of this initially cool, unsorted conglomerate by the decay of radioactive elements and the conversion of kinetic and potential energy to heat resulted in the development of Earth's liquid iron core and the gross internal zonation of the planet (i.e., differentiation into core, mantle, and crust). It has been concluded that Earth's core formed over a period of about 500 million years. It is likely that core formation resulted in the escape of an original primitive atmosphere and its replacement by one derived from the loss of volatile substances from the planetary interior.

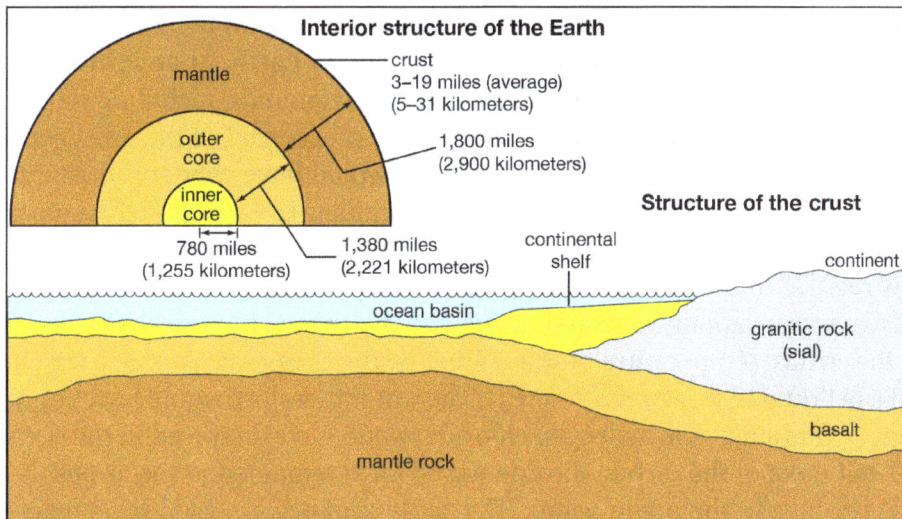

Earth's interior Earth's major structural layers displayed in cross section. Near-surface and surface features, such as ocean basins and continents, are also shown in cross section.

At an early stage, Earth thus did not have water or water vapour at its surface, and the topic of how water arrived on Earth's surface is a matter of substantial debate. Some scientists argue that much of the water on the planet was delivered by comet and meteor impacts. Both of these celestial objects have been shown to contain ice. Other scientists claim that most of Earth's water came from chemical reactions within the planet's interior. Once the planet's surface had cooled sufficiently, water contained in the minerals of the accreted material and released at depth could escape to the surface and, instead of being lost to space, cooled and condensed to form the initial hydrosphere. A large cool Earth most certainly served as a better trap for water than a small hot body because the lower the temperature, the less likelihood for water vapour to escape, and the larger the planetary mass, the stronger its gravitational attraction for water vapour. Whether most of the degassing took place during core formation or shortly thereafter or whether there has been significant degassing of Earth's interior throughout geologic time remains uncertain. It is likely that the hydrosphere attained its present volume early in Earth's history, and since that time there have been only small losses and gains. Gains would be from continuous degassing of Earth; the present degassing rate of juvenile water has been determined as being only 0.3 cubic km (about 0.07 cubic mile) per year. Water loss in the upper atmosphere is by photodissociation, the breakup of water vapour molecules into hydrogen and oxygen due to the energy of ultraviolet light. The hydrogen is lost to space and the oxygen remains behind. Only about 4.8×10^{-4} cubic km (about 0.0005 cubic mile) of water vapour is presently destroyed each year by photodissociation. This low rate can be readily explained: the very cold temperatures of the upper atmosphere result in a cold trap at an altitude of about 15 km (about 9.3 miles), where most of the water vapour condenses and returns to lower altitudes, thereby escaping photodissociation. Since the early formation of the hydrosphere, the amount of water vapour in the atmosphere has been regulated by the temperature of Earth's surface—hence its radiation balance. Higher temperatures imply higher concentrations of atmospheric water vapour, while lower temperatures suggest lower atmospheric levels.

Early Hydrosphere

The gases released from Earth during its early history, including water vapour, have been called excess volatiles because their masses cannot be accounted for simply by rock weathering. These volatiles are thought to have formed the early atmosphere of Earth. At an initial crustal temperature of about 600 °C (about 1,100 °F), almost all of these compounds, including water (H_2O), would have been in the atmosphere. The sequence of events that occurred as the crust cooled is difficult to reconstruct. Below 100 °C (212 °F) all of the water would have condensed, and the acid gases would have reacted with the original igneous crustal minerals to form sediments and an initial hydrosphere that was dominated by a salty ocean. If the chemical reaction rates are assumed to have been slow relative to cooling, an atmosphere of 600 °C would have contained, together with other compounds, water vapour, carbon dioxide, and hydrogen chloride (HCl) in a ratio of 20:3:1 and cooled to the critical temperature of water (that is, the temperature above which a compound cannot remain as a liquid—i.e., 374 °C [705 °F]). The water therefore would have condensed into an early hot ocean. At this stage, the hydrogen chloride would have dissolved in the ocean (about one mole per litre), but most of the carbon dioxide would have remained in the atmosphere, with only about 0.5 mole per litre in the ocean water. This early acid ocean would have reacted vigorously with crustal minerals, dissolving out silica and cations and creating a residue composed principally of aluminous clay minerals that would form the sediments of the early ocean basins.

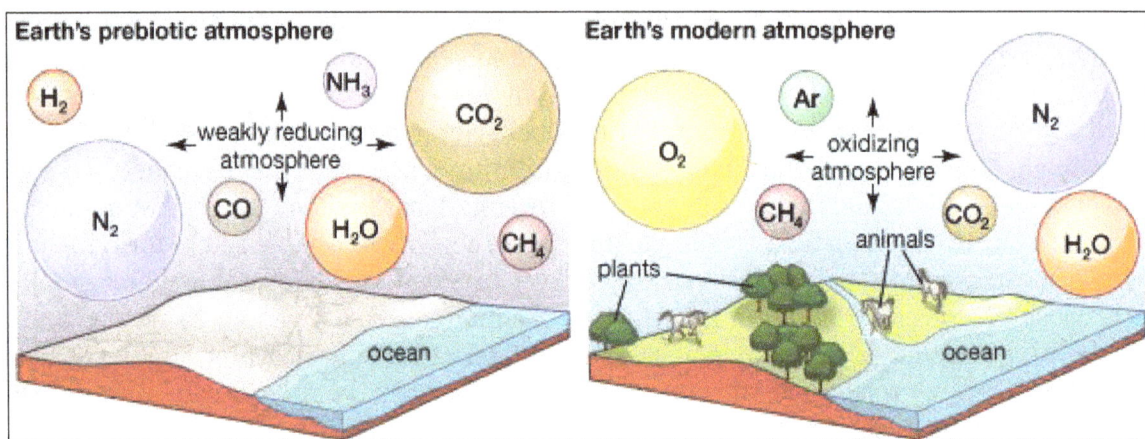

Earth's early and modern atmospheres.

Comparison of Earth's prebiotic and modern atmospheres. Before life began on the planet, Earth's atmosphere was largely made up of nitrogen and carbon dioxide gases. After photosynthesizing organisms multiplied on Earth's surface and in the oceans, much of the carbon dioxide was replaced with oxygen.

This is one of several possible pathways for the early surface of Earth. Whatever the actual case, after Earth's surface had cooled to 100 °C, it would have taken only a short time for the remaining acid gases to be consumed in reactions involving igneous rock minerals. The presence of cyanobacteria (e.g., blue-green algae) in the fossil record of rocks older than three billion years attests to the fact that Earth's surface had cooled to temperatures lower than 100 °C by this time, and neutralization of the original acid volatiles had taken place. It is possible, however, that, because of increased greenhouse gas concentrations early in the Archean Eon (4.0 billion to 2.5 billion years ago), Earth's surface could still have been warmer than today.

If most of the degassing of primary volatile substances from Earth's interior occurred early, the chloride released by the reaction of hydrochloric acid with rock minerals would be found in the oceans or in evaporite deposits, and the oceans would have a salinity and volume comparable to that of today. This conclusion is based on the assumption that there has been no drastic change in the ratios of volatiles released through geologic time. The overall generalized reaction indicative of the chemistry leading to the formation of the early oceans can be written in the form: primary igneous rock minerals + acid volatiles + H_2O → sedimentary rocks + oceans + atmosphere. It should be noted from this equation that, if all the acid volatiles and H_2O were released early in the history of Earth and in the proportions found today, then the total original sedimentary rock mass-produced would be equal to that of the present, and ocean salinity and volume would be close to those of today as well. If, on the other hand, degassing were linear with time, then the sedimentary rock mass would have accumulated at a linear rate, as would have oceanic volume. The salinity of the oceans, however, would remain nearly the same if the ratios of volatiles degassed did not change with time. The most likely situation is the one presented here—namely, that major degassing occurred early in Earth's history, after which minor amounts of volatiles were released episodically or continuously for the remainder of geologic time. The salt content of the oceans based on the constant proportions of volatiles released would depend primarily on the ratio of sodium chloride locked up in evaporites to that dissolved in the oceans. If all the sodium chloride in evaporites were added to the oceans today, the salinity would be approximately doubled. This value gives a sense of the maximum salinity that the oceans could have attained throughout geologic time.

One component absent from the early Earth's surface was free oxygen; it would not have been a constituent released from the cooling crust. Early production of oxygen was by the photodissociation of water in Earth's atmosphere, a process that was triggered by the absorption of the Sun's ultraviolet radiation. The reaction is $H_2O_{(liquid)}$ + hv → $H_{2\,(gas)}$ + $O_{2\,(gas)}$ in which hv represents the photon of ultraviolet light. The hydrogen produced would escape into space, while the oxygen would react with the early reduced gases by reactions such as $2H_2S + 3O_2$ → $2SO_2 + 2H_2O$. Oxygen production by photodissociation gave the early reduced atmosphere a start toward present-day conditions, but it was not until the appearance of photosynthetic organisms approximately three billion years ago that oxygen could accumulate in Earth's atmosphere at a rate sufficient to give rise to today's oxygenated environment. The photosynthetic reaction leading to oxygen production is given in equation,

$$106\,CO_3 + 16\,NO_3^- + HPO_4^{2-} + 122\,H_2O + 18\,H^+ \xrightarrow[\substack{\text{micro-}\\ \text{nutrients}}]{\substack{\text{radiant}\\ \text{energy}}} C_{106}H_{263}O_{110}N_{16}P + 138\,O_2 \,.$$

Transitional Hydrosphere

The nature of the rock record from the time of the first sedimentary rocks (approximately 3.8 billion years ago) to about one to two billion years ago suggests that the amount of oxygen in Earth's atmosphere was significantly lower than it is today and that there were continuous chemical trends in the sedimentary rocks formed and, more subtly, in the composition of the hydrosphere. The chemistry of rocks shifted dramatically during this transitional period. The source rocks of sediments during this time may have been more basaltic than subsequent ones. Sedimentary debris was formed by the alteration of such source rocks in an oxygen-deficient atmosphere and accumulated primarily under anaerobic (oxygen-free) marine conditions. The chief difference between reactions involving mineral-ocean equilibria at this time and the present day was the role played by ferrous iron (i.e., reduced state of iron). The concentration of dissolved iron in modern oceans is low because of the insolubility of oxidized iron oxides. During the transition stage and earlier, oxygen-deficient environments were prevalent, and these favoured the formation of minerals containing ferrous iron from the alteration of rocks slightly more rich in basalt than those of today. Indeed, iron carbonate siderite and iron silicate greenalite, in close association with chert and iron sulfide pyrite, are characteristic minerals that occur in iron formations of the middle Precambrian (about 2.5 billion to 1.6 billion years ago). The chert originally was deposited as amorphous silica; equilibrium between amorphous silica, siderite, and greenalite at 25 °C (77 °F) and a total pressure of one atmosphere requires a carbon dioxide pressure of about $10^{-2.5}$ (0.00316) atmosphere, or 10 times the present-day value.

A banded-iron formation (BIF) rock recovered from the Temagami greenstone belt, and dated to 2.7 billion years ago. Dark layers of iron oxide are intercalated with red chert.

The oceans of this transitional period can be thought of as a solution that resulted from an acid leach of basaltic rocks, and, because the neutralization of the volatile acid gases was not restricted primarily to land areas as it is today, much of this alteration may have occurred by submarine processes. Anaerobic depositional environments with internal carbon dioxide pressures of about $10^{-2.5}$ atmosphere prevailed, and the oxygen-deficient atmosphere itself may have had a carbon dioxide pressure close to $10^{-2.5}$ atmosphere. If so, the pH of early ocean water was lower than that of modern seawater and the calcium concentration was higher; moreover, the early ocean water was probably saturated with respect to amorphous silica—roughly 120 parts per million (ppm).

To simulate what might have occurred, it is helpful to imagine emptying the Pacific basin, throwing in great masses of broken basaltic material, filling it with hydrogen chloride dissolved in water so that the acid becomes neutralized, and then carbonating the solution by bubbling carbon dioxide through it. Oxygen would not be permitted into the system. The hydrochloric acid would leach the rocks, resulting in the release and precipitation of silica and the production of a chloride ocean containing sodium, potassium, calcium, magnesium, aluminum, iron, and reduced sulfur species in the proportions present in the rocks. As complete neutralization was approached, the aluminum could begin to precipitate as hydroxides and then combine with precipitated silica to form cation-deficient aluminosilicates. As the neutralization process reached its end, the aluminosilicates would combine with more silica and with cations to form such minerals as chlorite, and ferrous iron would combine with silica and sulfur to produce greenalite and pyrite. In the final solution, chlorine would be balanced by sodium and calcium in roughly equal proportions, with subordinate amounts of potassium and magnesium; aluminum would be quantitatively removed, and silicon would be at saturation with amorphous silica. If this solution were then carbonated, calcium would be removed as calcium carbonate, and the chlorine balance would be maintained by abstraction of more sodium from the primary rock. The sediments formed in this system would contain chiefly silica, ferrous iron silicates, chloritic minerals, calcium carbonate, calcium-magnesium carbonates, and small amounts of pyrite.

If the hydrogen chloride added were in excess of the carbon dioxide, the resultant oceans would have a high content of calcium chloride ($CaCl_2$), but with a pH still near neutrality. If the carbon dioxide added were in excess of the chlorine, calcium would be precipitated as carbonate until it reached a level roughly that of present-day ocean waters—namely, a few hundred parts per million.

If this newly created ocean were left undisturbed for several hundred million years, its waters would evaporate and be transported onto the continents (in the form of precipitation); streams would transport their loads into it. The sediments produced in this ocean would be uplifted and incorporated into the continents. The influence of the continental debris would gradually be felt and the pH might change somewhat. Iron would be oxidized out of the ferrous silicates to yield iron oxides, but the composition of the water would not vary substantially.

The primary minerals of igneous rocks are all mildly basic compounds. When these minerals react in excess with acids such as hydrogen chloride and carbon dioxide, they produce neutral or mildly alkaline solutions as well as a set of altered aluminosilicate and carbonate reaction products. It is improbable that seawater has changed through time from a solution approximately in equilibrium with these reaction products—i.e., with clay minerals and carbonates.

Modern Hydrosphere

Rain and dark cloud. Precipitation results when rising moisture-laden air cools and condenses into water droplets or ice crystals. When these structures become large enough to be acted upon by gravity, they fall to Earth.

It is likely that the hydrosphere achieved its modern chemical characteristics about 1.5 billion to 2 billion years ago. The chemical and mineralogical compositions and the relative proportions of sedimentary rocks of this age differ little from their counterparts of the Paleozoic Era (from 541 million to 252 million years ago). Calcium sulfate deposits dated to the late Precambrian (about 1.6 billion to 541 million years ago) attest to the fact that the acid sulfur gases had been neutralized to sulfate by this time. Chemically precipitated ferric oxides in late Precambrian sedimentary rocks indicate available free oxygen, whatever its percentage. The chemistry and mineralogy of middle and late Precambrian shales are similar to those of Paleozoic shales. The carbon isotopic signature of carbonate rocks has been remarkably constant for more than three billion years, indicating exceptional stability in size and fluxes related to organic carbon. The sulfur isotopic signature of sulfur phases in rocks strongly suggests that the sulfur cycle involving heterotrophic bacterial reduction of sulfate was in operation 2.7 billion years ago. It therefore appears that continuous cycling of sediments similar to those of today has occurred for 1.5 billion to 2 billion years and that these sediments have controlled hydrospheric, and particularly oceanic, composition.

It was once thought that the saltiness of the modern oceans simply represents the storage of salts (that is, compounds that form when part of an acid is replaced by a metal or something like a metal) derived from rock weathering and transported to the oceans by fluvial processes. With increasing knowledge of the age of Earth, however, it was soon realized that, at the present rate of delivery of salts to the ocean or even at much reduced rates, the total salt content and the mass of individual salts in the oceans could be attained in geologically short time intervals compared with the planet's age. The total mass of salt in the oceans can be accounted for at today's rates of stream delivery in about 12 million years. The mass of dissolved silica in ocean water can be doubled in just 20,000 years by the addition of stream-derived silica. To double the sodium content would take 70 million years. It then became apparent that the oceans are not merely an accumulator of salts. Rather, as water evaporates from the oceans, together with some salt, the salts introduced must be removed in the form of minerals deposited in marine sediments. Accordingly, the concept of the oceans as a chemical system changed from that of a simple accumulator to that of a steady-state system in which rates of inflow of materials equal rates of outflow. The steady-state concept permits influx to vary with time, but the inflow would be matched by nearly simultaneous and equal variation of efflux.

In such a steady-state conceptual view of the oceans, it has been found necessary to treat components of ocean water in terms of all their influxes and effluxes and to be more cognizant of the time scale of application of the steady-state concept. Indeed, the recent increase in the carbon dioxide concentration of the atmosphere due to the burning of fossil fuels has been linked to a changed pH and dissolved inorganic carbon concentrations of surface ocean water on a time scale measured in hundreds of years. If fossil-fuel burning were to cease, return to the original state of seawater composition could take thousands of years. Ocean water is not in steady state with respect to carbon on these time scales, but on a longer geologic time scale it certainly could be. Even on this longer time scale, however, oceanic composition has varied because of natural changes in the carbon dioxide level of the atmosphere and because of other factors.

It appears that the best description of modern seawater composition is that of a chemical system in a dynamic quasi-steady state. Changes in composition may occur over time, but the system always seems to return to a time-averaged steady-state composition. In other words, since 1.5 billion to 2 billion years ago, evolutionary chemical changes in the hydrosphere have been small when viewed against the magnitude of previous change.

It should be noted that rivers supply dissolved constituents to the oceans, whereas high- and low-temperature reactions between seawater and submarine basalts and reactions in sediment pore waters may add or remove constituents from ocean water. Biological processes involved in the formation of the opaline silica skeletons of diatoms (a type of algae with a shell made of silica) and radiolarians (planktonic protozoans) and the carbonate skeletons of planktonic foraminiferans (shelled unicellular organisms possessing pseudopods) and coccolithophorids (marine planktonic biflagellated organisms that secrete a type of minute calcium carbonate platelet or ring) chiefly remove calcium and silica from seawater. Exchange reactions between river-borne clays entering seawater are particularly significant for sodium and calcium ions. Most of the carbon imbalance in ocean water represents carbon released to the ocean-atmosphere system during precipitation of carbonate minerals; i.e.,

$$Ca^{2+} + 2\,HCO_3^- \rightarrow CaCO_3 + CO_2 + H_2O$$

In the case of iron, it has been documented that "dissolved" iron carried by rivers is rapidly precipitated as hydroxides in the mixing zone with seawater and that the reduced dissolved iron released from anaerobic sediments also is rapidly precipitated under the oxic conditions (i.e., those with oxygen present) prevailing in the water column. Iron is also precipitated as iron smectites, hydrated iron oxides, and nontronite (iron-rich montmorillonite) in the deep sea. It is thus likely that iron is removed by these processes.

Sodium Imbalance

The imbalance in sodium is large: 45 percent of the river input is not accounted for in the mass balance calculations. There are, however, major uncertainties in the estimation of the pore-water flux of sodium ions. An important sink for sodium on a geologic time scale is the formation of evaporites. If the amount of unbalanced sodium is expressed in terms of halite deposition, it would correspond to 1.6×10^{14} grams of sodium chloride per year as compared with a potential total depositional rate of 3.3×10^{14} grams annually. There are no important sodium chloride deposits forming today. Thus, one possibility is that sodium is accumulating in the oceans. If so, in 6×10^6 years at an accumulation rate of 63×10^{12} grams of sodium annually, the average salinity of the oceans would increase less than one part per thousand. The chlorine balance for the oceans, however, indicates that it is likely that the major problem in the imbalance for sodium lies in the flux estimates for sediment pore waters and perhaps submarine weathering processes.

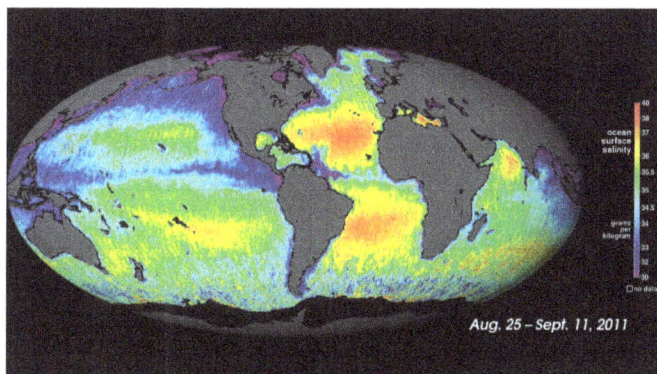

The first map of ocean salinity.

Modern seawater chemistry has been characteristic of roughly the past 600 million years of ocean history. Evaporite sediments provide strong evidence that the composition of seawater has not varied a great deal during this interval of geologic time. Nonetheless, it seems likely that fluctuations did occur, particularly in the concentrations of calcium, magnesium, and sulfate ions. The isotopic composition of sulfur in seawater, as recorded in evaporites, has varied dramatically during the past one billion years. Although it is difficult to relate these isotopic fluctuations to the calcium and sulfate concentrations of seawater, some scientists believe that the fluctuations do in fact imply changes in the latter. Furthermore, the major features of the sulfur isotopic curve for evaporites versus Phanerozoic time (the last 541 million years) is similar to that of the strontium-87/strontium-86 ratio, and perhaps the strontium/calcium ratio, of sedimentary materials during this time interval. Such covariation is consistent with a model in which fluxes related to alteration of seafloor basalts and continental river runoff vary with time, resulting in variation in seawater composition.

Changes in the Chemistry of the Atmosphere-hydrosphere

The chemistry of the atmosphere has certainly changed significantly during the past one billion years of Earth history. A modification of this kind implies changes in the chemistry of the hydrosphere as well. Oxygen in the atmosphere rose substantially between two billion years ago and the beginning of the Phanerozoic Eon (i.e., 541 million years ago), whereas atmospheric carbon dioxide levels probably decreased. This change led in general to a progressively more oxygenated and less acidic hydrosphere. It is likely that the development of higher land plants during the Devonian Period (from 419 million to about 359 million years ago) resulted in an increase in atmospheric oxygen and a decrease in carbon dioxide. Air trapped within bubbles in Arctic and Antarctic ice shows that the carbon dioxide content of the atmosphere during the climax of the last ice age was about 180 parts per million by volume (ppmv), and atmospheric CO_2 levels reached approximately 280 ppmv during the last great interglacial of 120,000 years ago, long before modern society initiated its extensive fossil-fuel burning and deforestation activities.

These atmospheric changes in themselves can influence the chemistry of the hydrosphere, but they also appear to be coupled with other changes in the rock-ocean-atmosphere-biota system that strongly affect hydrospheric chemistry. For example, though surface waters probably remained oxygenated during the Cretaceous and Devonian periods, there is evidence that intermediate and deep ocean waters were more anoxic (oxygen-depleted) than today. These "anoxic events" are characterized by dramatic changes in Earth's ocean-atmosphere system, including changes in the rates of the cyclic transfer of elements at Earth's surface and in atmospheric composition.

The probable impact of a bolide (either an asteroid or comet) and increased volcanism at the end of the Cretaceous (about 66 million years ago), though still the subject of hot debate, certainly could have caused short-term changes in the chemistry of Earth's atmosphere-hydrosphere. Scientists speculate that such an event could have had any of several results, including (1) changes in Earth's radiant-energy balance because of vast amounts of particulate and gaseous input into the atmosphere and subsequent cooling, (2) acid rain stemming from the input into the atmosphere of nitrogen gases generated by the bolide impact, (3) increased trace metal fluxes to the hydrosphere brought on by the destruction of the bolide and increased volcanism, and perhaps (4) increased "anoxia" of the hydrosphere as a result of the death of land and marine organisms.

Whatever the case, evidence of change is present in the rock record, albeit the composition of the modern hydrosphere has not varied greatly for the past one billion years. Moreover, as will be seen in the following section, humans are modifying the chemistry not only of local and regional water bodies but also of the entire global atmosphere-hydrosphere by increasing the rates of input of natural substances and by introducing new synthetic substances to the environment.

Impact of Human Activities on the Hydrosphere

The activities of modern society are having a severe impact on the hydrologic cycle. The dynamic steady state is being disturbed by the discharge of toxic chemicals, radioactive substances, and other industrial wastes and by the seepage of mineral fertilizers, herbicides, and pesticides into surface and subsurface aquatic systems. Inadvertent and deliberate discharge of petroleum, improper sewage disposal, and thermal pollution also are seriously affecting the quality of the hydrosphere.

The present discussion focuses on three major problems—eutrophication, acid rain, and the build-up of the so-called greenhouse gases. Each exemplifies human interference in the hydrologic cycle and its far-reaching effects.

Eutrophication

Historically, aquatic systems have been classified as oligotrophic or eutrophic. Oligotrophic waters are poorly fed by the nutrients nitrogen and phosphorus and have low concentrations of these constituents. There is thus low production of organic matter by photosynthesis in such waters. By contrast, eutrophic waters are well supplied with nutrients and generally have high concentrations of nitrogen and phosphorus and, correspondingly, large concentrations of plankton owing to high biological productivity. The waters of such aquatic systems are usually murky, and lakes and coastal marine systems may be oxygen-depleted at depth. The process of eutrophication is defined as high biological productivity resulting from increased input of nutrients or organic matter into aquatic systems. For lakes, this increased biological productivity usually leads to decreased lake volume because of the accumulation of organic detritus. Natural eutrophication occurs as aquatic systems fill in with organic matter; it is distinct from cultural eutrophication, which is caused by human intervention. The latter is characteristic of aquatic systems that have been artificially enriched by excess nutrients and organic matter from sewage, agriculture, and industry. Naturally eutrophic lakes may produce 75–250 grams of carbon per square metre per year, whereas those lakes experiencing eutrophication because of human activities can support 75–750 grams per square metre per year. Commonly, culturally eutrophic aquatic systems may exhibit extremely low oxygen concentrations in bottom waters. This is particularly true of stratified systems, as, for instance, lakes during summer where concentrations of molecular oxygen may reach levels of less than about one milligram per litre—a threshold for various biological and chemical processes.

Aquatic systems may change from oligotrophic to eutrophic, or the rate of eutrophication of a natural eutrophic system may be accelerated by the addition of nutrients and organic matter due to human activities. The process of cultural eutrophication, however, can be reversed if the excess nutrient and organic matter supply is shut off.

Not only do freshwater aquatic systems undergo cultural eutrophication, but coastal marine systems also may be affected by this process. On a global scale, the input by rivers of organic matter to the oceans today is twice the input in prehuman times, and the flux of nitrogen, together with that of phosphorus, has more than doubled. This excess loading of carbon, nitrogen, and phosphorus is leading to cultural eutrophication of marine systems. In several polluted eastern U.S. estuaries (e.g., Chesapeake and Delaware bays) and in some estuaries of western Europe (e.g., the Scheldt of Belgium and the Netherlands), all of the dissolved silica brought into the estuarine waters by rivers is removed by phytoplankton growth (primarily diatoms) resulting from excess fluxes of nutrients and organic matter. In the North Sea there is now a deficiency of silica and an excess of nitrogen and phosphorus, which in turn has led to a decrease in diatom productivity and an increase in cyanobacteria productivity—a biotic change brought about by cultural eutrophication.

Acid Rain

The emission of sulfur dioxide and nitrogen oxides to the atmosphere by human activities— primarily fossil-fuel burning—has led to the acidification of rain and freshwater aquatic systems. Acid rain is a worldwide problem and has been well documented for eastern North America and the countries of Western Europe.

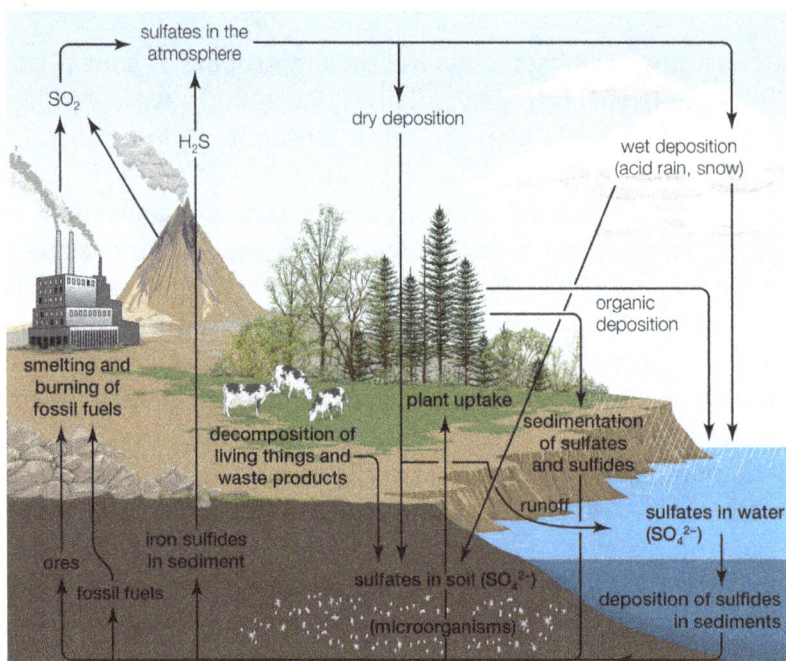

Sulfur cycle. Major sulfur-producing sources include sedimentary rocks, which release hydrogen sulfide gas, and human sources, such as smelters and fossil-fuel combustion, both of which release sulfur dioxide into the atmosphere.

Acid rain is defined as precipitation with a pH of less than 5.2 that results from reactions involving gases other than carbon dioxide. The overall reactions that produce such precipitation are those of equations $SO_2 + CH \rightarrow$ intermediate species $\rightarrow H_2SO_4$, $SO_2 + H_2O_2 = H_2SO_4.$, and $H_2SO_4 = 2H^+ + SO_4^{2-}$, and $NO_2 + OH \rightarrow HNO_3$.

Average pH can be calculated as the $-\log a_H^+$ (a_H^+ is activity of the hydrogen ion). A water chemistry study examining precipitation over the eastern United States for the period October 1979 through September 1980 revealed that low pH values were a result of equilibration of rainwater with the atmospheric acid gases of carbon, nitrogen, and sulfur. Equilibration only with atmospheric carbon dioxide would give a pH of 5.7. The significantly lower values are a result of reactions with nitrogen- and sulfur-bearing gaseous atmospheric components derived primarily from fossil-fuel burning sources.

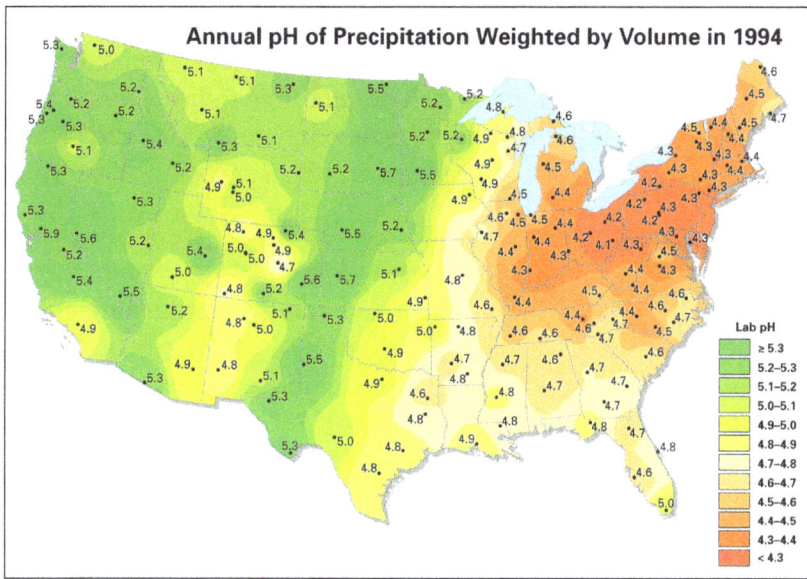

Annual pH of Precipitation Weighted by Volume in 1994

Nitrate and sulfate concentrations in precipitation over the eastern United States are strongly correlated with pH—the lower the pH of rain, the higher the concentrations of nitrate and sulfate. Such low pH values and increased nitrate and sulfate concentrations were observed in the rains of Western Europe and North America until the late 20th century. The pH values of precipitation in these regions have increased significantly since then because of strict air quality regulations. Other parts of the world that have industrialized since the late 20th century without enacting adequate air pollution controls, such as China, experienced similar pH declines in precipitation.

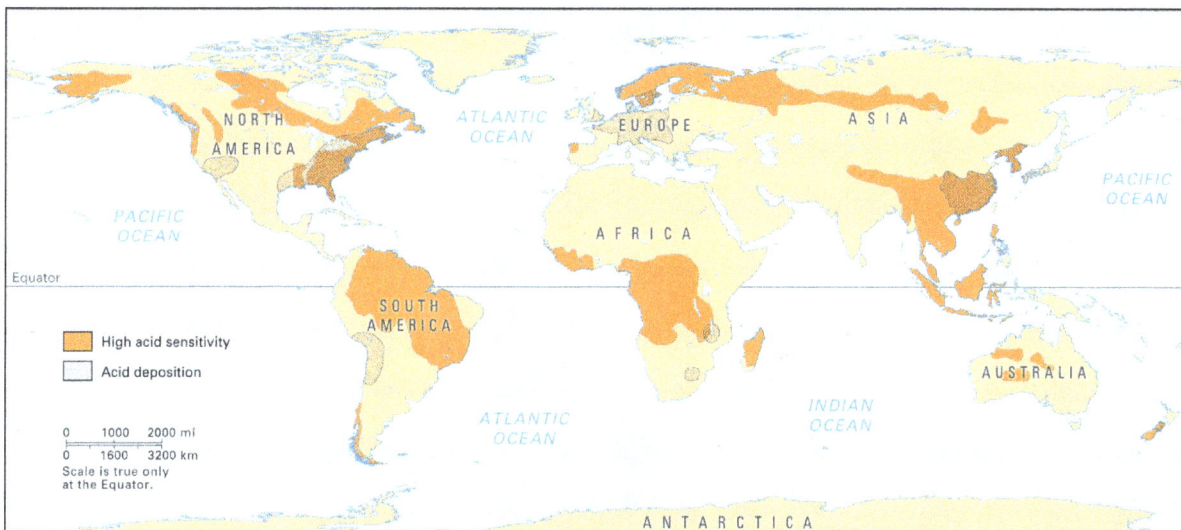

Areas affected by acid deposition contrasted with regions of high acid sensitivity.

Wet and dry deposition also removes the hydrogen ion produced in the rain by the oxidation and hydrolysis of these acid gases. This excess hydrogen ion can bring about the acidification of freshwater aquatic systems, particularly those with little buffer capacity (e.g., lakes situated in crystalline rock terrains). Furthermore, the lower pH values of rainwater, and consequently of soil water, can lead to increased mobilization of aluminum. Acidification of freshwater lakes in the eastern United States and southeastern Canada and increased aluminum concentrations in their waters are thought to be responsible for major changes in the ecosystems of the lakes. In particular, many lakes of this region lack substantial fish populations today, even though they supported large numbers of fish in the early 1900s. Acid rain also may be among the factors responsible for damage to the major forests and soils of the eastern United States and Western Europe.

Buildup of Greenhouse Gases

One problem that was brought about by human action and is definitely affecting the hydrosphere globally is that of the greenhouse gases (so called because of their heat-trapping "greenhouse" properties) emitted to the atmosphere. Of the greenhouse gases released by anthropogenic activities, carbon dioxide has received much attention. Measurements of carbon dioxide in air bubbles trapped in ice and the continuous measurement of carbon dioxide concentrations in air samples collected at Mauna Loa, Hawaii, since 1958 show that the atmospheric concentration of more than 400 ppmv is roughly 45 percent higher than its late 1700s value of 275 ppmv. Much of this increase is due to carbon dioxide released to the atmosphere from the burning of coal, oil, gas, and wood and from the slash-and-burn activities that accompany deforestation practices (as, for example, those adopted in the Amazon River basin). The component of the hydrosphere most greatly affected by this emission of carbon dioxide is the ocean.

Before human activities had substantially affected the carbon dioxide cycle, there was a net flux of carbon dioxide from the oceans through the atmosphere to the land, where the gas was used in the net production of organic matter and the chemical weathering of minerals in continental rocks. Because of fossil-fuel burning and land-use practices, the net transfer from the ocean to the land has been reversed, and the ocean has now become an important sink of carbon dioxide. The net chemical reaction of adding carbon dioxide to the ocean (provided there is no reaction with carbonate solids) is:

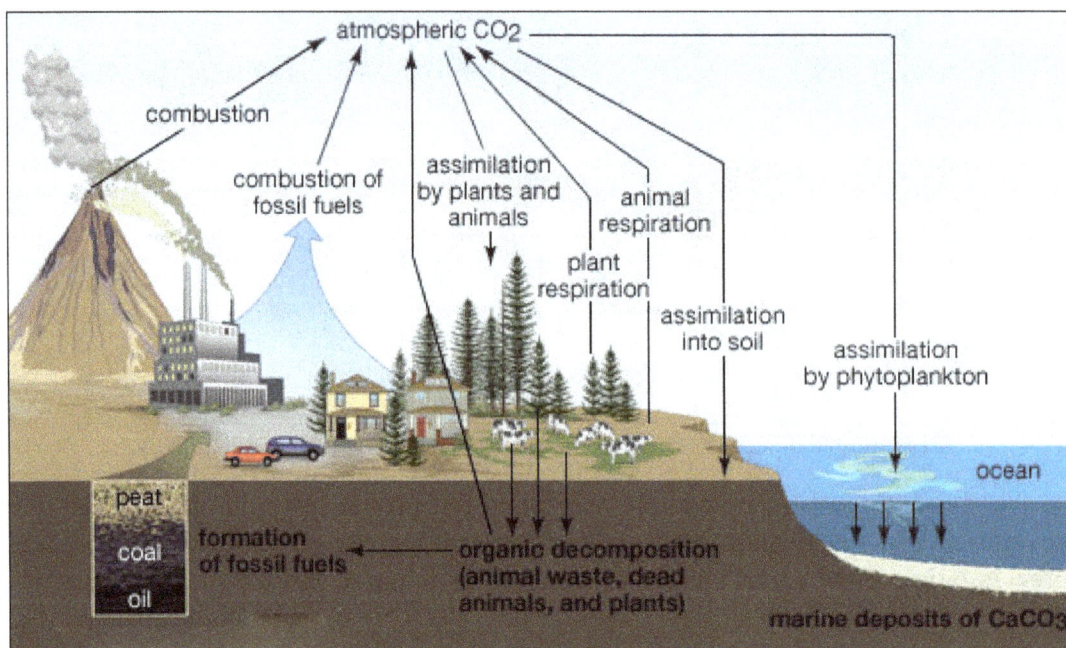

The generalized carbon cycle.

$$CO_2 + H_2O + CO_3^{2-}\ (carbonate\ ion) = 2\,HCO_3^-$$

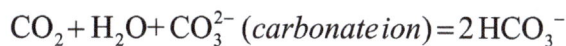

and a lowering of the pH of surface seawater. Such a pH effect has not been observed but conceivably could occur if carbon dioxide continues to be released to the atmosphere by human activities.

Based on greenhouse climate models and other considerations, it is possible that atmospheric carbon dioxide concentrations may double from their late 1700s level of 275 ppmv by the years 2030–50. By 2015, atmospheric carbon dioxide concentrations had surpassed 400 ppmv for the first time in 800,000 years. Climate models, which also consider the long-term warming potential made by other greenhouse gases (e.g., methane and nitrous oxide) in addition to that of carbon dioxide, project a rise in global mean surface temperature of 0.3 to 4.8 °C (0.5 to 8.6 °F) by 2100. This projected temperature increase would be two to three times greater at the poles than at the Equator and greater in the Arctic than in the Antarctic. At present there is no systematic worldwide program to decrease greenhouse gas emissions, except for that affecting chlorofluorocarbon (Freon) releases. Thus, it is conceivable that atmospheric carbon dioxide concentrations in the late 21st and early 22nd centuries might reach levels greater than twice their 1700s value. Whatever the case, the effect of the potential rise in surface temperature would be to speed up the hydrologic cycle and probably the rate of chemical weathering of continental rocks. Increases in the global mean evaporation and precipitation rates are expected from a doubling of the carbon dioxide level and a few degrees rise in global mean temperature. The effect on the water balance would be regional in nature, with some places becoming wetter and others drier. In general, there would be a trend toward greater and longer periods of summer dryness induced by lower soil moisture content and higher evaporation rates in the mid-latitudes of the Northern Hemisphere. In the arid western regions of the United States, which depend on irrigation for growing plants, severe water shortages could occur. By contrast, precipitation and runoff might increase, except in summer, at latitudes beyond 60° N because of a greater poleward transport of moisture. In summer, in a zone centred around 60° N, greater dryness might occur as a result of an earlier end of snowmelt.

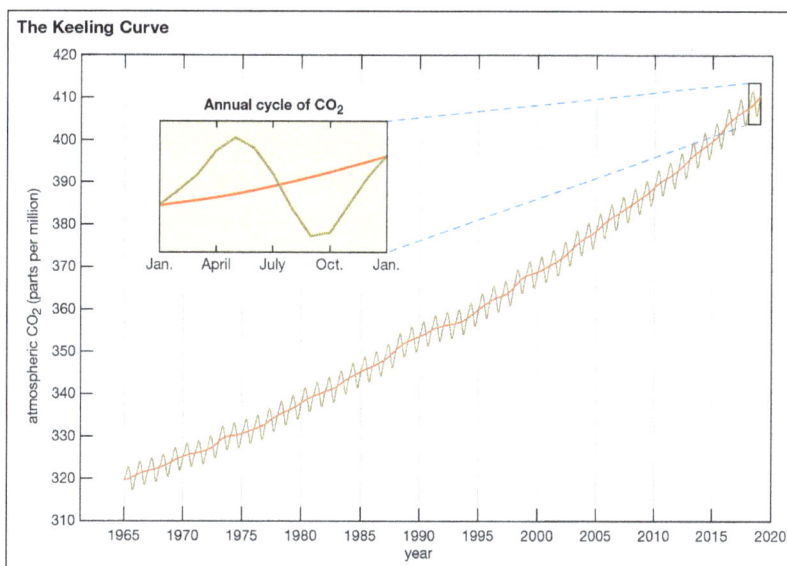

The Keeling Curve, named after American climate scientist Charles David Keeling, tracks changes in the concentration of carbon dioxide (CO_2) in Earth's atmosphere at a research station on Mauna Loa in Hawaii. Although these concentrations experience small seasonal fluctuations, the overall trend shows that CO_2 is increasing in the atmosphere.

Global warming could further affect the hydrologic cycle by the melting of ice and snow in the Greenland and Antarctic ice caps and in mountain glaciers, resulting in the transfer of water to the oceans. This process, together with thermal expansion of the oceans because of global warming, resulted in a slow rise in sea level during the 20th century, and it is expected to continue throughout the 21st century. If the West Antarctic Ice Sheet were to disintegrate, a much larger and more rapid rise in sea level of more than 3 metres (nearly 10 feet) could occur over the next several hundred years. The melting of all glacial ice would raise the sea level more than 66 metres (about 216 feet).

Significant reductions in the areal extent and thickness of sea ice in the Arctic occurring during the early 21st century have been attributed to global warming. Complete melting of the Arctic sea ice might occur, causing a northward shift in storm tracks and a reduction in Northern Hemispheric precipitation during the spring and fall. Furthermore, a worldwide reduction in sea ice might lead to increased evaporation from the ocean and increased low-altitude cloudiness, which would reflect solar radiation and cause cooling.

The potential changes in the hydrologic cycle induced by global warming resulting from anthropogenic emissions of greenhouse gases do not seem great. Yet, their consequences could be severe for ecosystems and human populations, especially since the latter are so sensitive to and dependent on such changes. A global rise in sea level of 1 metre (3.3 feet), for example, would almost completely inundate the coastal areas of Bangladesh. Agricultural lands could be displaced, just as patterns of arid, semiarid, and wet lands might become modified. It is essential that society plan for such potential changes so that, if they do occur, appropriate adjustments can be made to ameliorate them.

Chemistry of Water

Water is, in fact, a chemical. Its chemical formula is H_2O (or, less commonly, HOH), which is what this website is named after. H_2O is also one of the most well-known chemical formulas.

Each water molecule is made of two hydrogen atoms and one oxygen atom, thus there are two "H" atoms and one "O". The atoms are joined by covalent bonding, meaning that they share electrons (as opposed to ionic bonding, in which atoms completely transfer electrons). On the right is an image of water's molecular structure.

Water is the most abundant molecule on Earth. Approximately 70% of the Earth's surface is water. Water is also the only substance on Earth which naturally occurs in a solid, liquid and gas form.

The Celsius temperature scale is based on waters' freezing point (0 degrees) and boiling point (100 degrees). Unlike most substances, water's solid from is less dense than its liquid form - this is why ice cubes will float in your drinks. Water is also capable of absorbing a lot of heat before its temperature increases; thus it is used in things such as a radiator coolant in cars.

Water has a high surface tension. This is why some bugs like water striders (of the Gerridae family) can walk on water - because they weigh less than the surface tension of the water. Due to the shape of a water molecule, molecules stick and clump together to form this high tension (the two lighter hydrogen atoms in the image on the right could attach to the oxygen atom of another water molecule, and so forth). That's why water comes together in the form of drops - if it wasn't for gravity water would attach together in a spherical shape.

The chemistry of water deals with the fundamental chemical property and information about water. Water chemistry is discussed in the following subtitles.

- Composition of water;

- Structure and bonding of water;

- Molecular Vibration of water;

- Symmetry of water molecules;

- Formation of hydrogen bonding in water;

- Structure of ice;

- Autoionization;

- Leveling effect of water and acid-base characters;

- Amphiprotic nature;

- Reactivity of water towards alkali metals; alkaline earth metals; halogens; hydrides; methane; oxides; and oxygen ions;

- Electrolysis of water.

Physical Properties

Water is the chemical substance with chemical formula H_2O; one molecule of water has two hydrogen atoms covalently bonded to a single oxygen atom. Water is a tasteless, odorless liquid at ambient temperature and pressure. Liquid water has weak absorption bands at wavelengths of around 750 nm which cause it to appear to have a blue colour. This can easily be observed in a water-filled bath or wash-basin whose lining is white. Large ice crystals, as in glaciers, also appear blue.

Unlike other analogous hydrides of the oxygen family, water is primarily a liquid under standard conditions due to hydrogen bonding. The molecules of water are constantly moving in relation to each other, and the hydrogen bonds are continually breaking and reforming at timescales faster than 200 femtoseconds (2×10^{-13} seconds). However, these bonds are strong enough to create many of the peculiar properties of water, some of which make it integral to life.

Water, Ice and Vapor

Within the Earth's atmosphere and surface, the liquid phase is the most common and is the form that is generally denoted by the word "water". The solid phase of water is known as ice and commonly takes the structure of hard, amalgamated crystals, such as ice cubes, or loosely accumulated granular crystals, like snow. Aside from common hexagonal crystalline ice, other crystalline and amorphous phases of ice are known. The gaseous phase of water is known as water vapor (or steam). Visible steam and clouds are formed from minute droplets of water suspended in the air.

Water also forms a supercritical fluid. The critical temperature is 647 K and the critical pressure is 22.064 MPa. In nature this only rarely occurs in extremely hostile conditions. A likely example of naturally occurring supercritical water is in the hottest parts of deep water hydrothermal vents, in which water is heated to the critical temperature by volcanic plumes and the critical pressure is caused by the weight of the ocean at the extreme depths where the vents are located. This pressure is reached at a depth of about 2200 meters: much less than the mean depth of the ocean (3800 meters).

Heat Capacity and Heats of Vaporization and Fusion

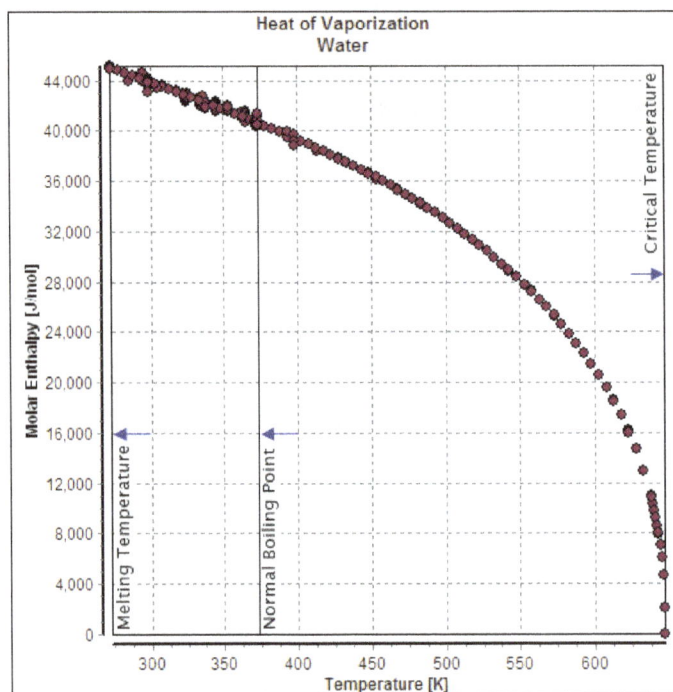

Heat of vaporization of water from melting to critical temperature.

Water has a very high specific heat capacity of 4.1814 J/(g·K) at 25 °C – the second highest among all the heteroatomic species (after ammonia), as well as a high heat of vaporization (40.65 kJ/mol or 2257 kJ/kg at the normal boiling point), both of which are a result of the extensive hydrogen bonding between its molecules. These two unusual properties allow water to moderate Earth's climate by buffering large fluctuations in temperature. Most of the additional energy stored in the climate system since 1970 has accumulated in the oceans.

The specific enthalpy of fusion (more commonly known as latent heat) of water is 333.55 kJ/kg at 0 °C: the same amount of energy is required to melt ice as to warm ice from −160 °C up to its melting point or to heat the same amount of water by about 80 °C. Of common substances, only that of ammonia is higher. This property confers resistance to melting on the ice of glaciers and drift ice. Before and since the advent of mechanical refrigeration, ice was and still is in common use for retarding food spoilage.

The specific heat capacity of ice at −10 °C is 2.03 J/(g·K) and the heat capacity of steam at 100 °C is 2.08 J/(g·K).

Density of Water and Ice

The density of water is about 1 gram per cubic centimetre (62 lb/cu ft): this relationship was originally used to define the gram. The density varies with temperature, but not linearly: as the temperature increases, the density rises to a peak at 3.98 °C (39.16 °F) and then decreases; this is unusual. Regular, hexagonal ice is also less dense than liquid water—upon freezing, the density of water decreases by about 9%.

These effects are due to the reduction of thermal motion with cooling, which allows water molecules to form more hydrogen bonds that prevent the molecules from coming close to each other. While below 4 °C the breakage of hydrogen bonds due to heating allows water molecules to pack closer despite the increase in the thermal motion (which tends to expand a liquid), above 4 °C water expands as the temperature increases. Water near the boiling point is about 4% less dense than water at 4 °C (39 °F).

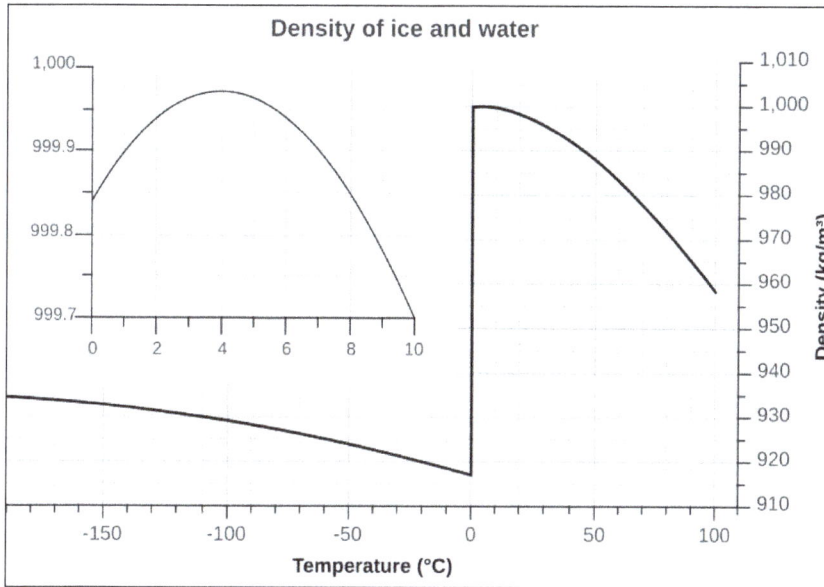

Density of ice and water as a function of temperature.

Under increasing pressure, ice undergoes a number of transitions to other polymorphs with higher density than liquid water, such as ice II, ice III, high-density amorphous ice (HDA), and very-high-density amorphous ice (VHDA).

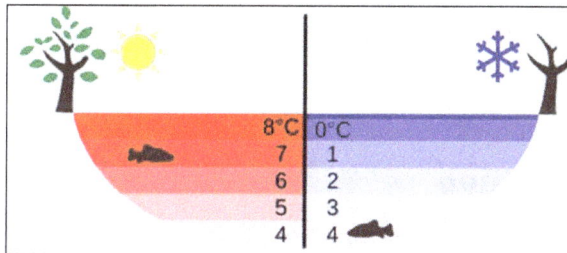

Temperature distribution in a lake in summer and winter.

The unusual density curve and lower density of ice than of water is vital to life—if water were most dense at the freezing point, then in winter the very cold water at the surface of lakes and other water bodies would sink, the lake could freeze from the bottom up, and all life in them would be killed. Furthermore, given that water is a good thermal insulator (due to its heat capacity), some frozen lakes might not completely thaw in summer. The layer of ice that floats on top insulates the water below. Water at about 4 °C (39 °F) also sinks to the bottom, thus keeping the temperature of the water at the bottom constant.

Density of Saltwater and Ice

The density of salt water depends on the dissolved salt content as well as the temperature. Ice still floats in the oceans, otherwise they would freeze from the bottom up. However, the salt content of oceans lowers the freezing point by about 1.9 °C and lowers the temperature of the density maximum of water to the former freezing point at 0 °C. This is why, in ocean water, the downward convection of colder water is *not* blocked by an expansion of water as it becomes colder near the freezing point. The oceans' cold water near the freezing point continues to sink. So creatures that live at the bottom of cold oceans like the Arctic Ocean generally live in water 4 °C colder than at the bottom of frozen-over fresh water lakes and rivers.

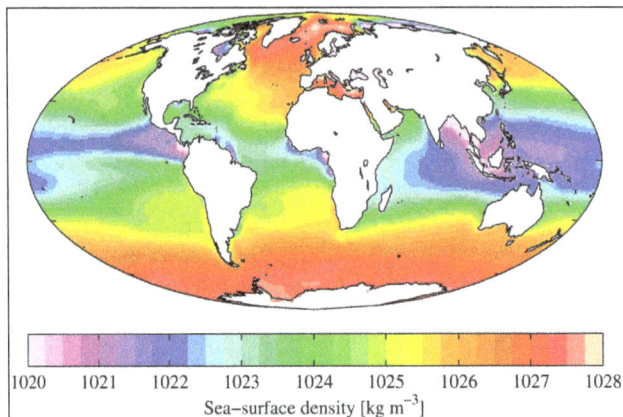

WOA surface density.

As the surface of salt water begins to freeze (at −1.9 °C for normal salinity seawater, 3.5%) the ice that forms is essentially salt-free, with about the same density as freshwater ice. This ice floats on the surface, and the salt that is "frozen out" adds to the salinity and density of the sea water just below it, in a process known as *brine rejection*. This denser salt water sinks by convection and the replacing seawater is subject to the same process. This produces essentially freshwater ice at −1.9 °C on the surface. The increased density of the sea water beneath the forming ice causes it to sink towards the bottom. On a large scale, the process of brine rejection and sinking cold salty water results in ocean currents forming to transport such water away from the Poles, leading to a global system of currents called the thermohaline circulation.

Miscibility and Condensation

Water is miscible with many liquids, including ethanol in all proportions. Water and most oils are immiscible usually forming layers according to increasing density from the top. This can be predicted by comparing the polarity. Water being a relatively polar compound will tend to be miscible with liquids of high polarity such as ethanol and acetone, whereas compounds with low polarity will tend to be immiscible and poorly soluble such as with hydrocarbons.

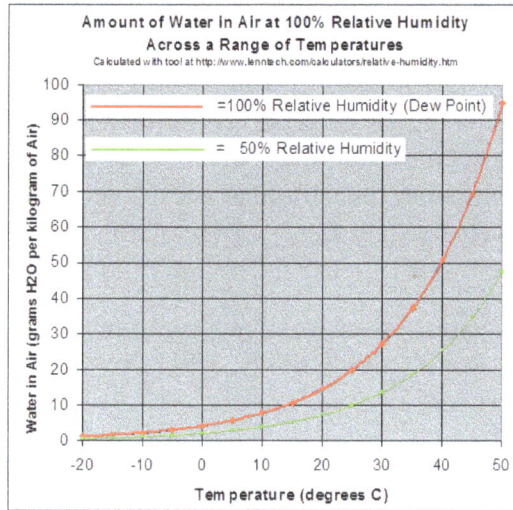

Amount of Water in Air at 100% Relative Humidity
Across a Range of Temperatures
Calculated with tool at http://www.lenntech.com/calculators/relative-humidity.htm

Red line shows saturation.

As a gas, water vapor is completely miscible with air. On the other hand, the maximum water vapor pressure that is thermodynamically stable with the liquid (or solid) at a given temperature is relatively low compared with total atmospheric pressure. For example, if the vapor's partial pressure is 2% of atmospheric pressure and the air is cooled from 25 °C, starting at about 22 °C water will start to condense, defining the dew point, and creating fog or dew. The reverse process accounts for the fog burning off in the morning. If the humidity is increased at room temperature, for example, by running a hot shower or a bath, and the temperature stays about the same, the vapor soon reaches the pressure for phase change, and then condenses out as minute water droplets, commonly referred to as steam.

A saturated gas or one with 100% relative humidity is when the vapor pressure of water in the air is at equilibrium with vapor pressure due to (liquid) water; water (or ice, if cool enough) will fail to lose mass through evaporation when exposed to saturated air. Because the amount of water vapor in air is small, relative humidity, the ratio of the partial pressure due to the water vapor to the saturated partial vapor pressure, is much more useful. Vapor pressure above 100% relative humidity is called super-saturated and can occur if air is rapidly cooled, for example, by rising suddenly in an updraft.

Vapor Pressure

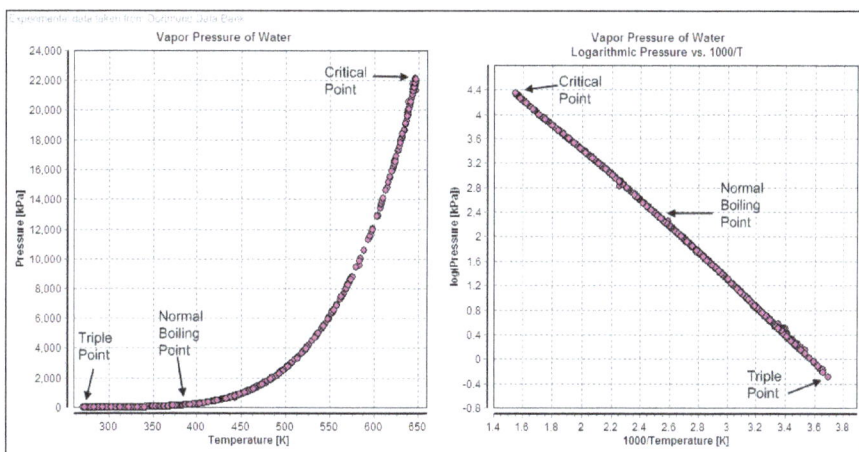

Vapor pressure diagrams of water.

Compressibility

The compressibility of water is a function of pressure and temperature. At 0 °C, at the limit of zero pressure, the compressibility is 5.1×10^{-10} Pa^{-1}. At the zero-pressure limit, the compressibility reaches a minimum of 4.4×10^{-10} Pa^{-1} around 45 °C before increasing again with increasing temperature. As the pressure is increased, the compressibility decreases, being 3.9×10^{-10} Pa^{-1} at 0 °C and 100 megapascals (1,000 bar).

The bulk modulus of water is about 2.2 GPa. The low compressibility of non-gases, and of water in particular, leads to their often being assumed as incompressible. The low compressibility of water means that even in the deep oceans at 4 km depth, where pressures are 40 MPa, there is only a 1.8% decrease in volume.

Triple Point

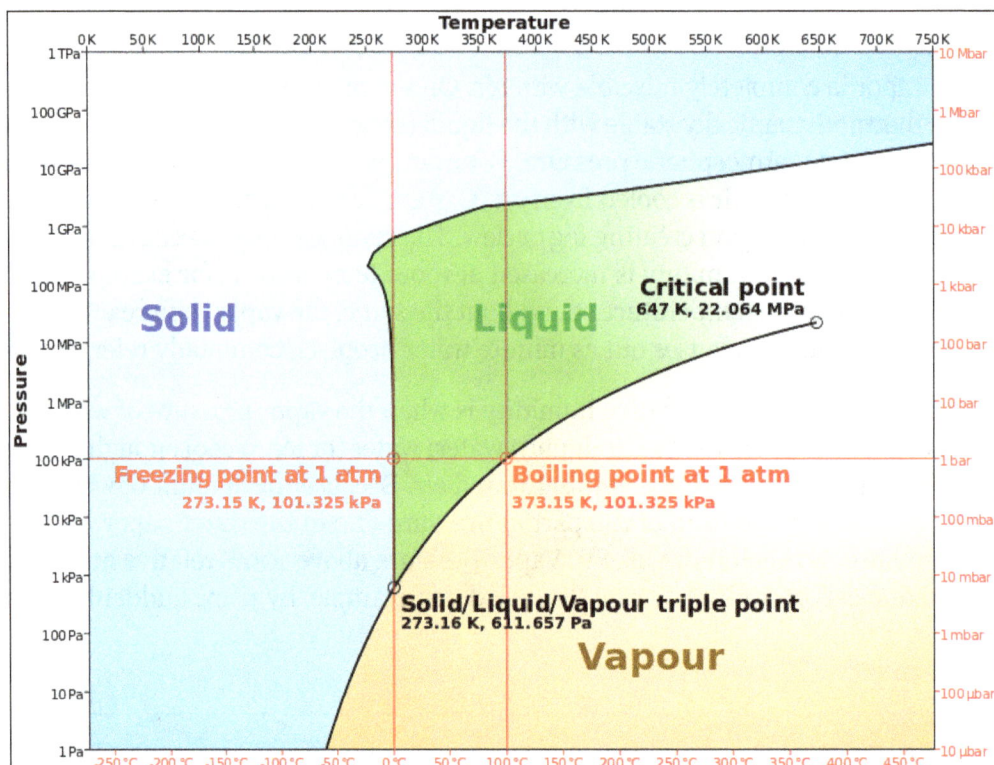

The Solid/Liquid/Vapour triple point of liquid water, ice I$_h$ and water vapor in the lower left portion of a water phase diagram.

The temperature and pressure at which ordinary solid, liquid, and gaseous water coexist in equilibrium is a triple point of water. Since 1954, this point had been used to define the base unit of temperature, the kelvin but, starting in 2019, the kelvin is now defined using the Boltzmann constant, rather than the triple point of water.

Due to the existence of many polymorphs (forms) of ice, water has other triple points, which have either three polymorphs of ice or two polymorphs of ice and liquid in equilibrium. Gustav Heinrich Johann Apollon Tammann in Göttingen produced data on several other triple points in the early 20th century. Kamb and others documented further triple points in the 1960s.

The various triple points of water		
Phases in stable equilibrium	Pressure	Temperature
liquid water, ice I_h, and water vapor	611.657 Pa	273.16 K (0.01 °C)
liquid water, ice I_h, and ice III	209.9 MPa	251 K (−22 °C)
liquid water, ice III, and ice V	350.1 MPa	−17.0 °C
liquid water, ice V, and ice VI	632.4 MPa	0.16 °C
ice I_h, Ice II, and ice III	213 MPa	−35 °C
ice II, ice III, and ice V	344 MPa	−24 °C
ice II, ice V, and ice VI	626 MPa	−70 °C

Melting Point

The melting point of ice is 0 °C (32 °F; 273 K) at standard pressure; however, pure liquid water can be supercooled well below that temperature without freezing if the liquid is not mechanically disturbed. It can remain in a fluid state down to its homogeneous nucleation point of about 231 K (−42 °C; −44 °F). The melting point of ordinary hexagonal ice falls slightly under moderately high pressures, by 0.0073 °C (0.0131 °F)/atm or about 0.5 °C (0.90 °F)/70 atm as the stabilization energy of hydrogen bonding is exceeded by intermolecular repulsion, but as ice transforms into its polymorphs above 209.9 MPa (2,072 atm), the melting point increases markedly with pressure, i.e., reaching 355 K (82 °C) at 2.216 GPa (21,870 atm) (triple point of Ice VII).

Electrical Properties

Electrical Conductivity

Pure water containing no exogenous ions is an excellent insulator, but not even "deionized" water is completely free of ions. Water undergoes auto-ionization in the liquid state, when two water molecules form one hydroxide anion (OH^-) and one hydronium cation (H_3O^+).

Because water is such a good solvent, it almost always has some solute dissolved in it, often a salt. If water has even a tiny amount of such an impurity, then the ions can carry charges back and forth, allowing the water to conduct electricity far more readily.

It is known that the theoretical maximum electrical resistivity for water is approximately 18.2 MΩ·cm (182 kΩ·m) at 25 °C. This figure agrees well with what is typically seen on reverse osmosis, ultra-filtered and deionized ultra-pure water systems used, for instance, in semiconductor manufacturing plants. A salt or acid contaminant level exceeding even 100 parts per trillion (ppt) in otherwise ultra-pure water begins to noticeably lower its resistivity by up to several kΩ·m.

In pure water, sensitive equipment can detect a very slight electrical conductivity of 0.05501 ± 0.0001 μS/cm at 25.00 °C. Water can also be electrolyzed into oxygen and hydrogen gases but in the absence of dissolved ions this is a very slow process, as very little current is conducted. In ice, the primary charge carriers are protons. Ice was previously thought to have a small but measurable conductivity of 1×10^{-10} S/cm, but this conductivity is now thought to be almost entirely from surface defects, and without those, ice is an insulator with an immeasurably small conductivity.

Polarity and Hydrogen Bonding

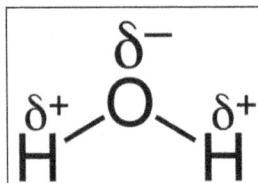

A diagram showing the partial charges
on the atoms in a water molecule.

An important feature of water is its polar nature. The structure has a bent molecular geometry for the two hydrogens from the oxygen vertex. The oxygen atom also has two lone pairs of electrons. One effect usually ascribed to the lone pairs is that the H–O–H gas phase bend angle is 104.48°, which is smaller than the typical tetrahedral angle of 109.47°. The lone pairs are closer to the oxygen atom than the electrons sigma bonded to the hydrogens, so they require more space. The increased repulsion of the lone pairs forces the O–H bonds closer to each other.

Another consequence of its structure is that water is a polar molecule. Due to the difference in electronegativity, a bond dipole moment points from each H to the O, making the oxygen partially negative and each hydrogen partially positive. A large molecular dipole, points from a region between the two hydrogen atoms to the oxygen atom. The charge differences cause water molecules to aggregate (the relatively positive areas being attracted to the relatively negative areas). This attraction, hydrogen bonding, explains many of the properties of water, such as its solvent properties.

Although hydrogen bonding is a relatively weak attraction compared to the covalent bonds within the water molecule itself, it is responsible for a number of water's physical properties. These properties include its relatively high melting and boiling point temperatures: more energy is required to break the hydrogen bonds between water molecules. In contrast, hydrogen sulfide (H_2S), has much weaker hydrogen bonding due to sulfur's lower electronegativity. H_2S is a gas at room temperature, in spite of hydrogen sulfide having nearly twice the molar mass of water. The extra bonding between water molecules also gives liquid water a large specific heat capacity. This high heat capacity makes water a good heat storage medium (coolant) and heat shield.

Cohesion and Adhesion

Water molecules stay close to each other (cohesion), due to the collective action of hydrogen bonds between water molecules. These hydrogen bonds are constantly breaking, with new bonds being formed with different water molecules; but at any given time in a sample of liquid water, a large portion of the molecules are held together by such bonds.

Water also has high adhesion properties because of its polar nature. On extremely clean/smooth glass the water may form a thin film because the molecular forces between glass and water molecules (adhesive forces) are stronger than the cohesive forces. In biological cells and organelles, water is in contact with membrane and protein surfaces that are hydrophilic; that is, surfaces that have a strong attraction to water. Irving Langmuir observed a strong repulsive force between hydrophilic surfaces. To dehydrate hydrophilic surfaces—to remove the strongly held layers of water of hydration—requires doing substantial work against these forces, called hydration forces. These forces are very large but decrease rapidly over a nanometer or less. They are important in biology, particularly when cells are dehydrated by exposure to dry atmospheres or to extracellular freezing.

Dew drops adhering to a spider web.

Surface Tension

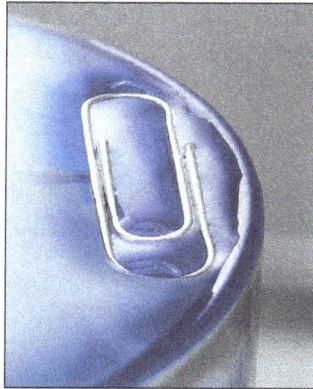

This paper clip is under the water level, which has risen gently and smoothly. Surface tension prevents the clip from submerging and the water from overflowing the glass edges.

Rain water flux from a canopy. Among the forces that govern drop formation: Surface tension, Cohesion (chemistry), Van der Waals force, Plateau–Rayleigh instability.

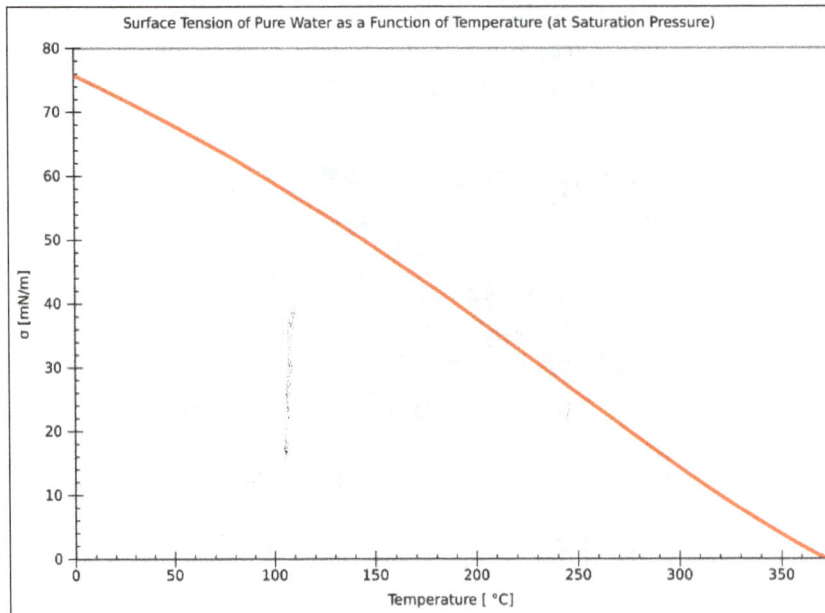

Temperature dependence of the surface tension of pure water.

Water has an unusually high surface tension of 71.99 mN/m at 25 °C which is caused by the strength of the hydrogen bonding between water molecules. This allows insects to walk on water.

Capillary Action

Because water has strong cohesive and adhesive forces, it exhibits capillary action. Strong cohesion from hydrogen bonding and adhesion allows trees to transport water more than 100 m upward.

Water as a Solvent

Presence of colloidal calcium carbonate from high concentrations of dissolved lime turns the water of turquoise.

Water is an excellent solvent due to its high dielectric constant. Substances that mix well and dissolve in water are known as hydrophilic ("water-loving") substances, while those that do not mix well with water are known as hydrophobic ("water-fearing") substances. The ability of a substance to dissolve in water is determined by whether or not the substance can match or better the strong attractive forces that water molecules generate between other water molecules. If a substance has properties that do not allow it to overcome these strong intermolecular forces, the molecules are precipitated out from the water. Contrary to the common misconception, water and hydrophobic substances do not "repel", and the hydration of a hydrophobic surface is energetically, but not entropically, favorable.

When an ionic or polar compound enters water, it is surrounded by water molecules (hydration). The relatively small size of water molecules (~ 3 angstroms) allows many water molecules to surround one molecule of solute. The partially negative dipole ends of the water are attracted to positively charged components of the solute, and vice versa for the positive dipole ends.

In general, ionic and polar substances such as acids, alcohols, and salts are relatively soluble in water, and non-polar substances such as fats and oils are not. Non-polar molecules stay together in water because it is energetically more favorable for the water molecules to hydrogen bond to each other than to engage in van der Waals interactions with non-polar molecules.

An example of an ionic solute is table salt; the sodium chloride, $NaCl$, separates into Na^+ cations and Cl^- anions, each being surrounded by water molecules. The ions are then easily transported away from their crystalline lattice into solution. An example of a nonionic solute is table sugar. The water dipoles make hydrogen bonds with the polar regions of the sugar molecule (OH groups) and allow it to be carried away into solution.

Quantum Tunneling

The quantum tunneling dynamics in water was reported as early as 1992. At that time it was known that there are motions which destroy and regenerate the weak hydrogen bond by internal rotations of the substituent water monomers. On 18 March 2016, it was reported that the hydrogen bond can be broken by quantum tunneling in the water hexamer. Unlike previously reported tunneling motions in water, this involved the concerted breaking of two hydrogen bonds. Later in the same year, the discovery of the quantum tunneling of water molecules was reported.

Electromagnetic Absorption

Water is relatively transparent to visible light, near ultraviolet light, and far-red light, but it absorbs most ultraviolet light, infrared light, and microwaves. Most photoreceptors and photosynthetic pigments utilize the portion of the light spectrum that is transmitted well through water. Microwave ovens take advantage of water's opacity to microwave radiation to heat the water inside of foods. Water's light blue colour is caused by weak absorption in the red part of the visible spectrum.

Structure

A single water molecule can participate in a maximum of four hydrogen bonds because it can accept two bonds using the lone pairs on oxygen and donate two hydrogen atoms. Other molecules like hydrogen fluoride, ammonia and methanol can also form hydrogen bonds. However, they do not show anomalous thermodynamic, kinetic or structural properties like those observed in water because none of them can form four hydrogen bonds: either they cannot donate or accept hydrogen atoms, or there are steric effects in bulky residues. In water, intermolecular tetrahedral structures form due to the four hydrogen bonds, thereby forming an open structure and a three-dimensional bonding network, resulting in the anomalous decrease in density when cooled below 4 °C. This repeated, constantly reorganizing unit defines a three-dimensional network extending throughout the liquid. This view is based upon neutron scattering studies and computer simulations, and it makes sense in the light of the unambiguously tetrahedral arrangement of water molecules in ice structures.

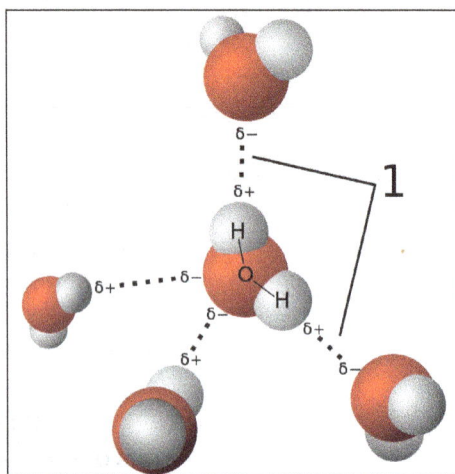

Model of hydrogen bonds (1) between molecules of water.

However, there is an alternative theory for the structure of water. In 2004, a controversial paper from Stockholm University suggested that water molecules in liquid form typically bind not to four but to only two others; thus forming chains and rings. The term "string theory of water" was coined. These observations were based upon X-ray absorption spectroscopy that probed the local environment of individual oxygen atoms.

Molecular Structure

The repulsive effects of the two lone pairs on the oxygen atom cause water to have a bent, not linear, molecular structure, allowing it to be polar. The hydrogen-oxygen-hydrogen angle is 104.45°, which is less than the 109.47° for ideal sp^3 hybridization. The valence bond theory explanation is that the oxygen atom's lone pairs are physically larger and therefore take up more space than the oxygen atom's bonds to the hydrogen atoms. The molecular orbital theory explanation (Bent's rule) is that lowering the energy of the oxygen atom's nonbonding hybrid orbitals (by assigning them more s character and less p character) and correspondingly raising the energy of the oxygen atom's hybrid orbitals bonded to the hydrogen atoms (by assigning them more p character and less s character) has the net effect of lowering the energy of the occupied molecular orbitals because the energy of the oxygen atom's nonbonding hybrid orbitals contributes completely to the energy of the oxygen atom's lone pairs while the energy of the oxygen atom's other two hybrid orbitals contributes only partially to the energy of the bonding orbitals (the remainder of the contribution coming from the hydrogen atoms' 1s orbitals).

Isotopologues

Several isotopes of both hydrogen and oxygen exist, giving rise to several known isotopologues of water. Vienna Standard Mean Ocean Water is the current international standard for water isotopes. Naturally occurring water is almost completely composed of the neutron-less hydrogen isotope protium. Only 155 ppm include deuterium (^{2}H or D), a hydrogen isotope with one neutron, and fewer than 20 parts per quintillion include tritium (^{3}H or T), which has two neutrons. Oxygen also has three stable isotopes, with ^{16}O present in 99.76%, ^{17}O in 0.04%, and ^{18}O in 0.2% of water molecules.

Deuterium oxide, D_2O, is also known as heavy water because of its higher density. It is used in nuclear reactors as a neutron moderator. Tritium is radioactive, decaying with a half-life of 4500 days; THO exists in nature only in minute quantities, being produced primarily via cosmic ray-induced nuclear reactions in the atmosphere. Water with one protium and one deuterium atom HDO occurs naturally in ordinary water in low concentrations (~0.03%) and D_2O in far lower amounts (0.000003%) and any such molecules are temporary as the atoms recombine.

The most notable physical differences between H_2O and D_2O, other than the simple difference in specific mass, involve properties that are affected by hydrogen bonding, such as freezing and boiling, and other kinetic effects. This is because the nucleus of deuterium is twice as heavy as protium, and this causes noticeable differences in bonding energies. The difference in boiling points allows the isotopologues to be separated. The self-diffusion coefficient of H_2O at 25 °C is 23% higher than the value of D_2O. Because water molecules exchange hydrogen atoms with one another, hydrogen deuterium oxide (DOH) is much more common in low-purity heavy water than pure dideuterium monoxide D_2O.

Consumption of pure isolated D_2O may affect biochemical processes – ingestion of large amounts impairs kidney and central nervous system function. Small quantities can be consumed without any ill-effects; humans are generally unaware of taste differences, but sometimes report a burning sensation or sweet flavor. Very large amounts of heavy water must be consumed for any toxicity to become apparent. Rats, however, are able to avoid heavy water by smell, and it is toxic to many animals.

Light water refers to deuterium-depleted water (DDW), water in which the deuterium content has been reduced below the standard 155 ppm level.

Occurrence

Water is the most abundant substance on Earth and also the third most abundant molecule in the universe, after H_2 and CO. 0.23 ppm of the earth's mass is water and 97.39% of the global water volume of 1.38×10^9 km³ is found in the oceans.

Reactions

Acid-base Reactions

Water is amphoteric: it has the ability to act as either an acid or a base in chemical reactions. According to the Brønsted-Lowry definition, an acid is a proton (H^+) donor and a base is a proton acceptor. When reacting with a stronger acid, water acts as a base; when reacting with a stronger base, it acts as an acid. For instance, water receives an H^+ion from HCl when hydrochloric acid is formed:

$$HCl(acid) + H_2O(base) \rightleftharpoons H_3O^+ + Cl^-$$

In the reaction with ammonia, NH_3, water donates a H^+ ion, and is thus acting as an acid:

$$NH_3(base) + H_2O(acid) \rightleftharpoons NH_4^+ + OH^-$$

Because the oxygen atom in water has two lone pairs, water often acts as a Lewis base, or electron pair donor, in reactions with Lewis acids, although it can also react with Lewis bases, forming hydrogen bonds between the electron pair donors and the hydrogen atoms of water. HSAB theory describes water as both a weak hard acid and a weak hard base, meaning that it reacts preferentially with other hard species:

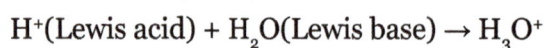

$$H^+(\text{Lewis acid}) + H_2O(\text{Lewis base}) \rightarrow H_3O^+$$

$$Fe^{3+}(\text{Lewis acid}) + H_2O(\text{Lewis base}) \rightarrow Fe(H_2O)^{3+}_6$$

$$Cl^-(\text{Lewis base}) + H_2O(\text{Lewis acid}) \rightarrow Cl(H_2O)^-_6$$

When a salt of a weak acid or of a weak base is dissolved in water, water can partially hydrolyze the salt, producing the corresponding base or acid, which gives aqueous solutions of soap and baking soda their basic pH:

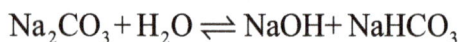

$$Na_2CO_3 + H_2O \rightleftharpoons NaOH + NaHCO_3$$

Ligand Chemistry

Water's Lewis base character makes it a common ligand in transition metal complexes, examples of which include metal aquo complexes such as $Fe(H_2O)^{2+}_6$ to perrhenic acid, which contains two water molecules coordinated to a rhenium center. In solid hydrates, water can be either a ligand or simply lodged in the framework, or both. Thus, $FeSO_4 \cdot 7H_2O$ consists of $[Fe_2(H_2O)_6]^{2+}$ centers and one "lattice water". Water is typically a monodentate ligand, i.e., it forms only one bond with the central atom.

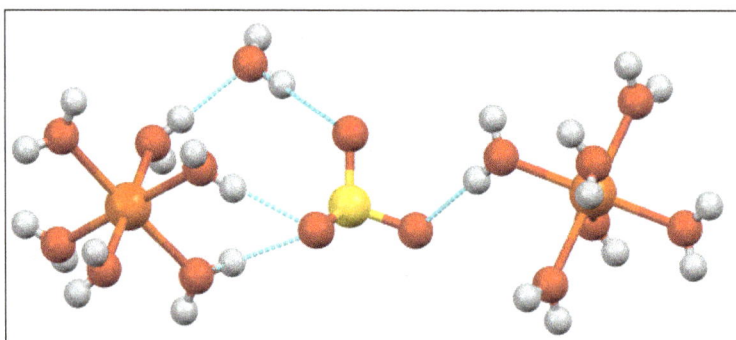

Some hydrogen-bonding contacts in $FeSO_4 \cdot 7H_2O$. This metal aquo complex crystallizes with one molecule of "lattice" water, which interacts with the sulfate and with the $[Fe(H_2O)_6]^{2+}$ centers.

Organic Chemistry

As a hard base, water reacts readily with organic carbocations; for example in a hydration reaction, a hydroxyl group (OH^-) and an acidic proton are added to the two carbon atoms bonded together in the carbon-carbon double bond, resulting in an alcohol. When addition of water to an organic molecule cleaves the molecule in two, hydrolysis is said to occur. Notable examples of hydrolysis are the saponification of fats and the digestion of proteins and polysaccharides. Water can also be a leaving group in $S_N 2$ substitution and E2 elimination reactions; the latter is then known as a dehydration reaction.

Water in Redox Reactions

Water contains hydrogen in the oxidation state +1 and oxygen in the oxidation state −2. It oxidizes chemicals such as hydrides, alkali metals, and some alkaline earth metals. One example of an alkali metal reacting with water is:

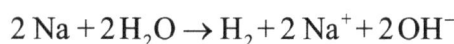

$$2\,Na + 2\,H_2O \rightarrow H_2 + 2\,Na^+ + 2\,OH^-$$

Some other reactive metals, such as aluminum and beryllium, are oxidized by water as well, but their oxides adhere to the metal and form a passive protective layer. Note that the rusting of iron is a reaction between iron and oxygen that is dissolved in water, not between iron and water.

Water can be oxidized to emit oxygen gas, but very few oxidants react with water even if their reduction potential is greater than the potential of O_2/H_2O. Almost all such reactions require a catalyst. An example of the oxidation of water is:

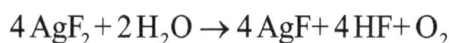

$$4\,AgF_2 + 2\,H_2O \rightarrow 4\,AgF + 4\,HF + O_2$$

Electrolysis

Water can be split into its constituent elements, hydrogen and oxygen, by passing an electric current through it. This process is called electrolysis. The cathode half reaction is:

$$2\,H^+ + 2\,e^- \rightarrow H_2$$

The anode half reaction is:

$$2\,H_2O \rightarrow O_2 + 4\,H^+ + 4\,e^-$$

The gases produced bubble to the surface, where they can be collected or ignited with a flame above the water if this was the intention. The required potential for the electrolysis of pure water is 1.23 V at 25 °C. The operating potential is actually 1.48 V or higher in practical electrolysis.

Chemical Properties

At standard conditions, water is a polar liquid that slightly dissociates disproportionately into a hydronium ion and hydroxide ion.

$$2\,H_2O \rightleftharpoons H_3O^+ + OH^-$$

The ionic product of pure water, K_w has a value of about 10–14 at 25 °C. Pure water has a concentration of the hydroxide ion (OH^-) equal to that of the hydrogen ion (H^+), which gives a pH of 7 at 25 °C.

Geochemistry

Action of water on rock over long periods of time typically leads to weathering and water erosion, physical processes that convert solid rocks and minerals into soil and sediment, but under some conditions chemical reactions with water occur as well, resulting in metasomatism or mineral hydration, a type of chemical alteration of a rock which produces clay minerals. It also occurs when Portland cement hardens.

Water ice can form clathrate compounds, known as clathrate hydrates, with a variety of small molecules that can be embedded in its spacious crystal lattice. The most notable of these is methane clathrate, $4CH_4 \cdot 23H_2O$, naturally found in large quantities on the ocean floor.

Acidity in Nature

Rain is generally mildly acidic, with a pH between 5.2 and 5.8 if not having any acid stronger than carbon dioxide. If high amounts of nitrogen and sulfur oxides are present in the air, they too will dissolve into the cloud and rain drops, producing acid rain.

pH

Pure distilled water is not a homogeneous mixture of H_2O molecules. Water molecules tend to ionize, that is to separate into H^+ and OH^- ions, which mix with the water. Normally, there are relatively few of these ions, and it is customary to measure the number of H^+ ions present to get a value known as *pH*. In pure H_2O at room temperature (20° C), there are about 0.0000001 grams of H^+ ions per liter (and an equivalent amount of OH^- ions). You can see that $0.0000001 = 1 \times 10^{-7}$, and, dredging up memories of mathematics long forgotten, you may even recall that the logarithm of this number is the exponent (-7). Make that exponent positive (take the negative exponent) and you have 7, the pH of pure water at room temperature.

Acids are substances which decrease the pH of a solution. They work by splitting up water molecules, moving in with the OH^-, and leaving "orphan" H^+ floating around. *Bases* work the opposite way; they split up the H_2O, but they combine with the H^+ and leave extra OH^- in solution. Of course, there are also those acids and bases which affect pH directly; hydrochloric acid (HCl) in H_2O disassociates to form H^+ and Cl^-, and sodium hydroxide (NaOH) disassociates to form Na^+ and OH^-. Acids and bases differ in their strength, that is, their ability to change pH.

To measure the effect of an acid or base, all we have to do is measure the H^+ ion concentration of the water. If the addition of a substance to the water increases the H^+ ion concentration, the substance is an acid; if it reduces the H^+ ion concentration, it is a base. Measurement of the H^+ ion concentration usually takes advantage of its electrical properties, and the most accurate determinations of pH are now made with electronic devices known as pH meters which use electrodes to measure the electrical properties of a solution under examination. If the solution has 0.000001 grams of H^+ ions per liter, this corresponds to a pH of 6 (an acidic solution); if there are 0.00000001 gr H^+/l then the pH is 8 (a basic solution).

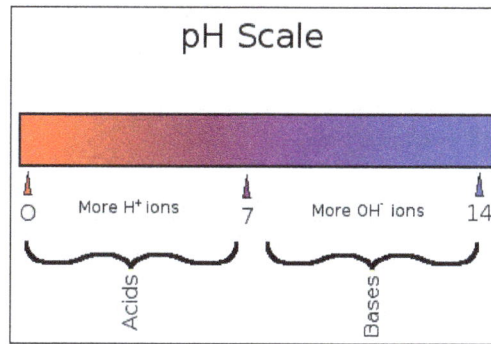

The pH scale, showing concentrations of both H⁺ and OH⁻ ions. Note that as pH increases,
the concentration of H⁺ ions decreases. Note also the logarithmic scale.

Note that, a change of one pH unit means a tenfold increase or decrease in the number of ions present; a change of two pH units means a hundredfold change in number, and so on. One of the reasons people do not worry about acid rain as much as they should is that they don't realize that small changes in pH numbers mean huge changes in the actual numbers of ions in the water.

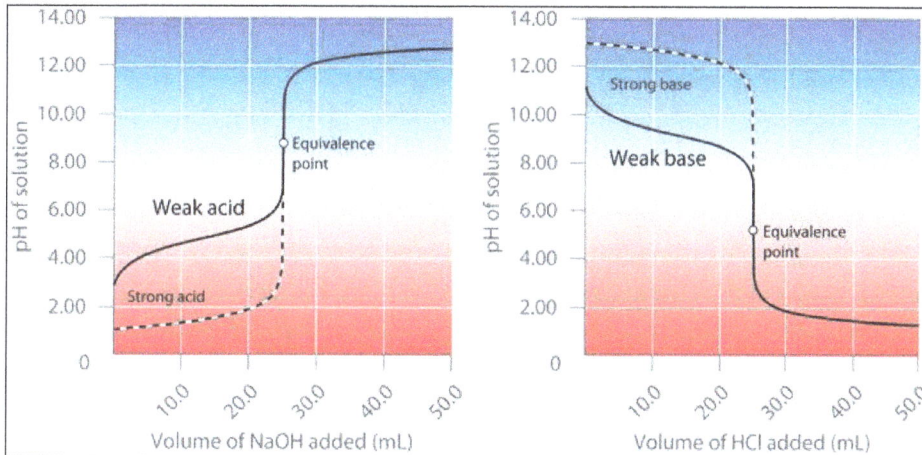

A portion of the pH scale with a natural y-axis; note the dramatic decrease
in H⁺ ions that occurs with each unit change in pH.

Most organisms operate best when the pH surrounding their cells is about 7, and natural waters usually oblige them. The pH of clean mountain lakes (almost pure water) is usually about 7; the pH of the oceans is about 8. Local conditions can affect pH. Addition of CO_2 to water decreases pH (makes the water acidic); removal of CO_2 makes water basic. Respiration and photosynthesis can therefore have an effect on pH, particularly in areas where water flow is negligible.

Buffers are compounds which reduce the effects of acids or bases. The presence of a buffer usually prevents major changes in pH even if a strong acid or base is added to the system. In the human body there are numerous buffer systems to prevent pH changes; these include proteins and inorganic buffers. In natural systems, all of the buffers are inorganic. Sodium bicarbonate ($NaHCO_3$) is one of the major natural buffers; it is abundant in the ocean, and in freshwater where there are limestone rocks (which are largely carbonate deposited from earlier, buffered oceans). Areas such as the Adirondacks, with mostly granite rocks and little limestone, have little buffering capacity and thus are susceptible to pH changes due to acid rain. The "buffers" in things like aspirin (an acid) prevent stomach pH from becoming even more acidic.

The pH of a solution is critical since it changes a number of parameters. Some substances change their solubility or reactivity as pH changes. Proteins may change their shape and thus change their reactivity. Almost anything that dissolves in H_2O will affect the pH somewhat, and we will consider those effects as we examine those substances.

Salinity

When water falls to earth, it is chemically pure - well, at least it used to be. In any event, it normally carries relatively few dissolved chemicals. We say that it has a low *salinity*. As it flows towards the oceans, it dissolves more and more molecules of various sorts from the rocks and soil it passes over - a process we call *weathering*. It also picks up materials released from plants and animals, either as waste products or through decay of their bodies. Since water does not leave the oceans except by evaporation (which removes only water, leaving the other materials behind), these materials accumulate in the oceans, and the oceans have a salinity much greater than that of freshwater. In turn, the salinity of freshwater increases as it flows to the oceans, and coastal areas such as *estuaries* may have a salinity reflecting the mixture of fresh and salt water. Other areas that allow water to leave only by evaporation also may also have salt water. For instance, if you do not periodically allow water to flow completely through your houseplant pots, or if you never remove water from your freshwater aquarium, they will become salty.

We call it salt water because most of the dissolved substances are just that - salts, simple ionic compounds that typically disassociate completely into positive and negative ions in water. Since an equal number of positive and negative ions are added in such cases, most salts do not change the pH appreciably, unless one of the components happens to be H^+ or OH^-. Note, however, that the presence of salts may affect the chemistry of certain pH tests, which may only be accurate when used in either fresh or salt water (some chemical tests will work in both). The most abundant of the salts found in the oceans is NaCl, sodium chloride or table salt. We measure salinity in terms of the number of grams of dissolved salts in 1000 g (one l) of seawater. Since all these salts affect the electrical conductivity of the water, it is simple to measure salinity with an electrical meter. The more current that passes through a solution, the greater its salinity. Seawater ranges in salinity, but a useful approximation is 35g/ kg; or 35 parts per thousand or 3.5%. Places like the Great Salt Lake, certain tidal pools, etc., can have higher salinities; most freshwater systems have dramatically lower salinities.

In seawater, the ions are dominated by Cl^- (19.353 gr/kg), Na^+ (10.76 gr/kg), SO_4^{-2} (2.712 gr/kg), Mg^{+2} (1.294 gr/kg), Ca^{+2} (0.413 gr/kg), K^+ (0.387 gr/kg), HCO_3^- (0.142 gr/kg), Br^- (0.067 gr/kg), and Sr^{+2} (0.008 gr/kg); other ions are present in trace amounts, including gold. Freshwater is dominated by similar ions, but the amounts are highly variable and depend on season, amount of rainfall, type of rocks, etc. Freshwater with an abundance of calcium and magnesium is termed *hard*; it resists forming suds when detergents are added and leaves a residue. Water softening agents often attempt to chelate these ions out of solution.

The salinity of water is important to organisms for two main reasons: osmoticity and density. We have already considered density; remember that salt water is more dense than freshwater, thus, it is easier to float in salt water. This also means that freshwater flowing into the ocean tends to remain on the surface for some time, and, further, that it takes some energy (tides, wind, currents) to get the two to mix effectively. It is not uncommon in estuaries for freshwater to lie on top of salt-water in layers. Temperature still plays a role; warm freshwater is very likely to lay on top of cold seawater, but other possibilities may also occur, such as warm seawater laying on top of a layer of cold freshwater. We will take up the concept of osmoticity next.

Osmotic Relations

Imagine water of two different salinities. We say that the more saline solution is *hypertonic* (or hyperosmotic) in relation to the other, or that the less saline solution is *hypotonic* (or hypoosmotic) in relation to the other. If the salinities were the same, they would be *isotonic*(isosmotic). Osmoticity, then, is simply a comparison of the salinities (or, more accurately, the number of dissolved particles, including non-ionic compounds) of two solutions. It is important to remember that, in considering osmoticity, that you must have two different solutions to compare - it is a relative term.

If you mix the two solutions, they will achieve an intermediate salinity. The process of *diffusion* means that all of the ions will, as a result of random chance, distribute themselves evenly throughout the solution. This occurs because the molecules and ions are in constant motion (except at 0 K); diffusion occurs more rapidly at higher temperatures. Stirring, mixing, bulk flow due to density differences (as a result of temperature or salinity) etc., will all decrease the time it takes for the solutions to come into equilibrium, with the dissolved material equally distributed.

Now, place the two solutions in contact with each other, but separated by a membrane. The membrane will have the property of allowing only small, uncharged (polar is ok too) molecules to pass through. Say the one solution is in a sack made of the membrane, and the sack is dropped into the other solution. Will diffusion take place?

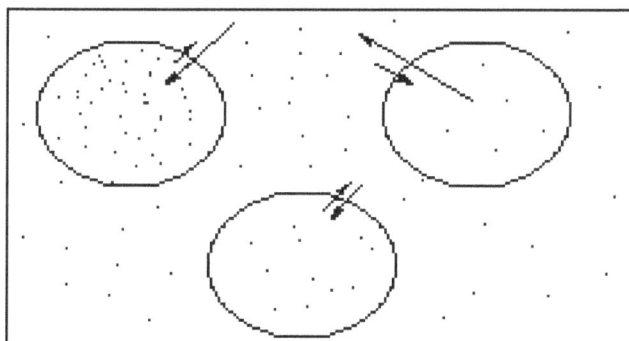

Diagrams of three hypothetical cells of varying osmoticity in relation to the surrounding medium. The dots indicate osmotic particles, the length of the arrows represents relative water flow in the direction of the arrow.

The answer is yes and no. Water molecules, being small and (relatively) uncharged, pass through easily, and in both directions. Ions get stopped at the border due to their charge, as do larger molecules such as sugars and proteins, regardless of their charges. For physical reasons we need not get into here, the water molecules will, on average, tend to go more to the hyperosmotic side of the membrane, although a few rugged individualists will still cross over from the hypertonic side to the hypotonic side. The *net* flow of water will be to the hypertonic side. If the outside solution in our example is hypertonic, water will flow out of the sack and the sack will shrink; if the outside solution is hypotonic, water will flow into the bag and it will swell and (perhaps) burst. We have just built a model of a cell(ions can penetrate the membrane of a cell, however).

All organisms are faced with the basic problem of maintaining a proper amount of water in their cells. There are two solutions; one is elegant in the extreme - allow water to move into or out of the cell as it pleases, and waste no energy on correcting it. This strategy is known as osmoconformation, and works only if the organism is in an unchanging environment - like the middle of the ocean, or deep inside a body. It is no accident that the salinity (though not the precise ionic make-up) of our cells - and the cells of most living things - is isotonic with seawater.

Moving into freshwater (or moving onto land) requires a different strategy. Freshwater is simply too dilute to keep life in a cell going. You need more stuff in your cells, and that stuff increases tonicity. Freshwater organisms face a constant influx of water from the surrounding hypotonic medium, and they can potentially lose important ions to that solution also. Therefore, the strategy among most freshwater organisms is to cover as much of the body as possible with an impermeable coat, and leave all water exchange to a relatively small number of cells. These cells will maintain the water balance, and the remaining cells are bathed in an isotonic solution. Cells can maintain osmotic balance by using ATP to actively pump Cl^- ions into the cell. The inside of the cell becomes negatively charged, and other ions, such as Na^+ come in because of this. Water that flows in is passed on to the blood and excreted as a dilute urine.

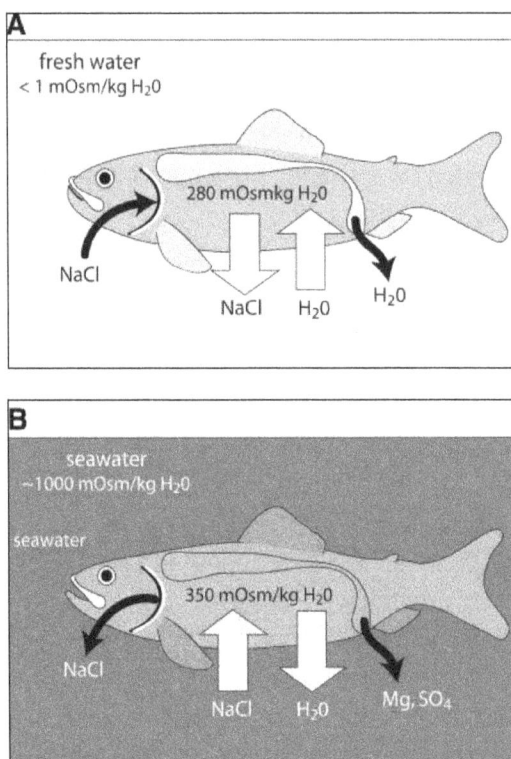

Diagrams of two hypothetical "fish", one in freshwater and one in saltwater.
Main sites of ion exchange are the gills and the excretory organs (kidneys).

Osmotic exchange also takes place across the lining of the gut. The freshwater fish gains water but loses ions passively across the gills; to compensate, the gills actively pump in ions and the kidneys form a dilute urine. The saltwater fish gains ions and loses water across the gills; to compensate, water is ingested (along with salt), the gills actively pump ions out of the body, and a small amount of relatively concentrated urine is formed.

Many marine organisms (those which maintain their internal fluids hypotonic to seawater, perhaps reflecting the salinity of the sea when they evolved), and all terrestrial organisms, face the problem of water loss. This is conquered in reverse fashion. These organisms drink seawater, absorb water (and ions) from the gut, and pump ions out through specialized cells.

The specialized cells in marine organisms are the same as those used by freshwater organisms to pump ions in; it's just that the membrane-bound proteins which form the ion "pumps" are "installed" backwards. These Cl⁻ pumping cells, whichever way they pump, are called *chloride cells*. In fish (both freshwater and marine), they are located on the gills. Because respiratory structures must have relatively permeable surfaces for gas exchange, they are also a common place to put chloride cells on a body which is otherwise impervious to water flow. Another popular place is in the gut and kidneys, in both places ion concentrations are manipulated to get water to flow where the organism wants it to.

Dissolved Gasses

The atmosphere is about 78% nitrogen, 21% oxygen, and only 0.033% carbon dioxide (but we're working on that). These three gasses have different solubilities in H_2O, however. To measure solubility, imagine this experiment: Replace all the air in the atmosphere with only the gas you are interested in so that at ground level the pressure is 1 atmosphere. If you now take one liter of water (let's say at 10° C), it will become saturated with the gas. In our imaginary world, if the gas was nitrogen, the water would hold 18.61 ml of it; if the gas was oxygen, the water would hold 38.46 ml of it; if the gas was carbon dioxide, the water would hold a whopping 1.194 ml of it. If you're wondering how 1 liter of water can hold 1.194 liters of gas, get a liter of Perrier and shake it up - then open it. The solubility experiment tells us that oxygen is about twice as soluble as nitrogen, and that CO_2 is about 1,000 times more soluble than either nitrogen or oxygen. Of course, in the real world there is not 1 atmosphere of pressure of each of these gasses, each gas has a *partial pressure* corresponding to the percentage of the atmosphere that it makes up. Multiplying the solubility of the gas by its percentage in the atmosphere gives the amount of that gas that will be dissolved in the water:

Solubility of Gasses in H_2O at 10° C

Gas	Percent	Solubility*	In water*
Nitrogen	78.084%	18.61	14.53
Oxygen	20.946%	38.46	8.06
Carbon Dioxide	0.033%	1,194.00	0.39
* Solubility in ml/l			

The absolute amount of a gas in water solution will vary with several factors. Increasing temperature will reduce the amount of gas that water can hold; you are familiar with this fact already, since it is manifested whenever you heat water (the small bubbles that form beforethe water boils). Decreasing pressure (increased altitude) will also decrease the amount of gas dissolved. Increasing salinity also decreases the ability of water to dissolve gasses; seawater holds about 20% less gas than freshwater, and hypersaline water holds even less gas. And, of course, there are other gasses which are dissolved in water besides these three (which are the major ones).

Oxygen

Oxygen, of course, is of critical importance to living organisms. It is important to remember that oxygen is a potent poison, and that too much can be a bad thing. We're probably lucky that the atmosphere has only 21% O_2; astronomers on other planets have probably concluded that life cannot exist on Earth because of the high levels of noxious oxygen. Any more O_2 and it would probably start to poison the process of photosynthesis (actually, it already does to a point, the reaction is called photorespiration). In water, however, it is easy for O_2 concentrations to become depleted in local areas, such as in sediments or the bottom of stagnant bodies of water. These local areas of oxygen depletion serve as important refugia for the many ecologically significant anaerobic organisms, and thus should not be viewed automatically as "bad". Nitrogen fixation, for instance, is notoriously sensitive to O_2; root nodules of legumes and heterocysts of Cyanobacteria are structures built to exclude O_2 so that nitrogen fixation can take place.

The solubility of O_2 in water is slightly more complex than was explained initially above. Better estimates of O_2 solubility involve moist air as opposed to the dry air in our theoretical planet experiment. Oxygen solubility is also very temperature dependent. In general, the solubility of oxygen can be estimated by the equation:

$$\text{Solubility} \left(mg \, / \, l \right) = \frac{468}{31.6 + t}$$

Where t is degrees C; note that we are now talking about milligrams, not milliliters. Thus, for a t of 10° C, there should be 11.25 mg O_2/l; at 20° C there should be 9.069 mg O_2/l; and at 0° C there should be 14.81 mg O_2/l at saturation. These values are for sea level and fresh water; at altitude or in salt water the saturation values would be lower. Water is not always at saturation, however. Organisms within the water can quickly use up dissolved O_2, and replacement by diffusion may not equal the rate of respiration, thus resulting in lower O_2 levels. Oxygen enters the water through diffusion from the atmosphere and from photosynthesis. Diffusion from the atmosphere may be enhanced by any type of turbulence such as water flowing over rocks, waves, wind, etc.; photosynthesis is dependent on many factors. Oxygen loss is primarily due to respiration of animals and plants in the water; respiration is temperature dependent, with more respiration occurring at higher temperatures.

These relationships set the stage for determining the critical factors for O_2 availability in water. More oxygen will be available in cool, sunny, turbulent habitats (uptake of O_2 by organisms is also enhanced by flow or turbulence since more water will pass over respiratory structures per unit time). Oxygen will be in short supply under warm, dark, stagnant conditions, where respiration is high and no photosynthesis is taking place. In natural aquatic systems, O_2 stress most often occurs at the water-sediment interface on dark, calm nights in the middle of the summer or dry season, when water flow is at a minimum. Any pollutant which adds to the amount of plant life present (fertilizers), or which contributes organic matter which will be decomposed by aerobic bacteria (sewage), will aggravate the O_2 stress at that time. Also in natural systems, the most O_2 is available in cool habitats with shallow water, particularly in turbulent headwater streams or near wave-swept coasts.

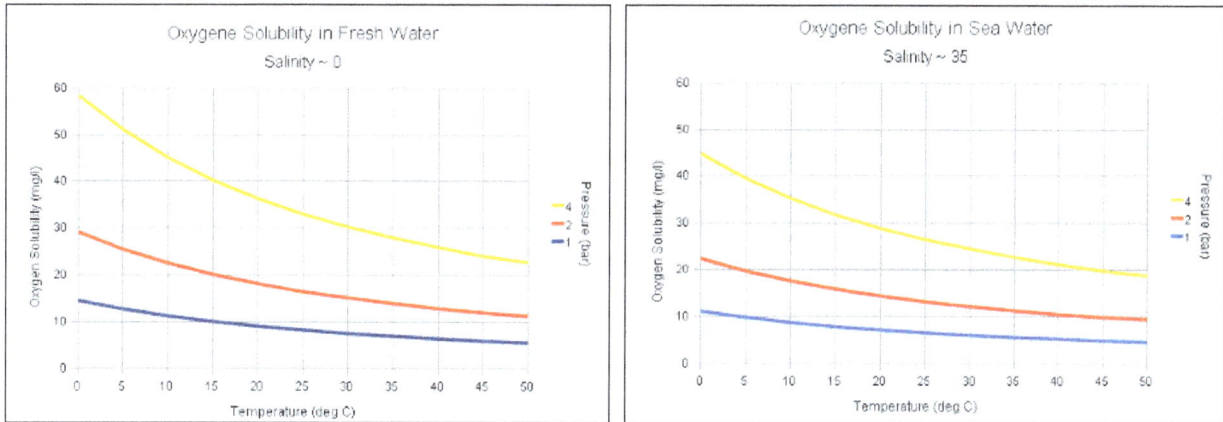

Oxygen solubility at saturation for fresh water at sea level at various temperatures. O_2 solubility decreases with increasing temperature, note also the single point plotted for salt water; salinity significantly decreases the amount of O_2 in solution.

Oxygen concentration in the water can be measured by either electrical meters or by chemical methods. The meters use electrodes and must be calibrated or adjusted to compensate for temperature (most include an electrical thermometer attached to the probe), altitude, and salinity. Chemical tests, such as the well-known *Winkler Method*, which measures oxygen by determining how much of a standard chemical is oxidized, work by titration. Results are often expressed in *percent saturation*; for instance, if the water is at 20° C and measures 8 mg O_2/l rather than the saturation value of 9.069 mg O_2/l, the percent saturation would be 88%. It is not unknown to have *supersaturated* water; this results in percent saturations over 100%, under such conditions O_2 is released to the atmosphere - which is how it got there in the first place.

Another important concept to consider is Biochemical Oxygen Demand, or *BOD*. BOD is the measure of how much oxygen is taken up from the water by both biological agents (organisms) and simple chemical reactions (like Fe + O_2 = rust). To measure BOD, a sample of the water (and/or sediment) is measured for O_2 content, sealed up, and left for a specified period (often 24 hours). The O_2 level is measured again, and the amount of O_2 used up (mg), or the rate of O_2 use (mg/hour) is reported as the BOD. Regulations governing sewage plants, stockyards, etc., may specify the amount of BOD that can be released each day.

Carbon Dioxide

As one is evolved in respiration or photosynthesis, the other is taken up. Still, the much greater solubility of CO_2 compared to O_2 means that the relationship of CO_2 to water is much different compared to that of O_2 and water. For one, CO_2 dissolved in H_2O increases acidity. Secondly, CO_2 can combine with other chemicals in the water to form different compounds; these compounds further affect the chemistry of the water. The primary sources of CO_2 in water are biological (respiration), geologic (weathering of carbonate containing rock such as limestone) and from the atmosphere. Rainwater in particular, with its high surface area, picks up a lot of CO_2 as it falls.

A small amount of CO_2 entering water combines with the water to form carbonic acid:

$$CO_2 + H_2O \leftrightarrow H_2CO_3$$

The carbonic acid, in turn, may dissociate:

$$H_2CO_3 + \leftrightarrow HCO_3 + H$$

The HCO_3- is known as *bicarbonate*, and may be taken up by photosynthesizing plants. Changing the pH of a solution changes the relative concentration of the various CO_2 forms; a decrease in pH (more acid) leads to more CO_2 and $H_2CO_3^-$; an increase in pH (more basic) leads to an increase of the CO_3^{-2} ion. In order to measure the amount of CO_2 present, one must obviously know the pH of the solution.

Carbon dioxide may also combine with water and metals such as magnesium and calcium to form other bicarbonates. The amount of CO_2 so combined is referred to as *alkalinity*, which really has nothing to do with OH^- concentration, but much to do with the buffering capacity of the water. It works like this: Highly alkaline water tends to have a high (basic) pH and will turn a phenolphthalein solution pink. If you add acid to it, the bicarbonates, with their negative charge, attract and bind the positive H^+ ions, and form carbonic acid. If you keep adding acid, eventually the pH changes to 8.3, and the pink fades. The amount of acid added corresponds to the *phenolphthalein alkalinity*, but not all the bicarbonate is converted at this point; in fact, it is at its peak. If you now add methyl orange, a dye that will change color at pH 4.4, and continue to add acid, you will drive more bicarbonate to form carbonic acid, which in turn reaches its peak at a pH of 4.4. The total amount of acid added thus corresponds to the amount of CO_2 present in the sample. This method works only if there are not significant numbers of non-carbonate negative ions to absorb H^+ ions.

Testing for CO_2 is usually done chemically, and is quite complicated. It is based on the above reactions and works *something* like this: if phenolphthalein added to a water solution turns pink, then the pH is over 8.3 and significant bicarbonates, including those of Ca^{++} and Mg^{++}, are present. The solution is titrated with acid to a pH of 4.4, and the total amount of acid added corresponds to the amount of CO_2 present (as bicarbonate). If the initial solution with phenolphthalein does not turn pink, then a lesser amount of CO_2 (as bicarbonate or whatever) is present, and bicarbonate is titrated until a pink color appears. The amount of bicarbonate added will then correspond inversely to the amount of CO_2 that was present initially; the more bicarbonate you have to add, the less was there to start with.

An interesting reaction may take place when CO_2 is removed from water during photosynthesis - a precipitate of $CaCO_3$ (calcium carbonate) may form. This reaction, along with calcium carbonate formation by corals, has lead to the production of most of the limestone in the world. Weathering of that limestone, in turn, yields much of the carbonate present in freshwaters, and, as we have seen, that carbonate forms an effective buffer against decreased pH due to acid rain. As a result of the many forms it can take, CO_2 is usually present in ample amounts for photosynthesis to occur in aquatic habitats. Although rarely a limiting factor, it is known that more alkaline lakes are, up to a point, more productive than less alkaline lakes. Other factors, such as the abundance of phosphorous and nitrogen, are more often limiting factors, and levels of these plant nutrients are usually correlated with alkalinity, making the independent factors hard to sort out.

Plant Nutrients: Nitrogen and Phosphorous

Aside from O_2 and CO_2, there are a variety of other chemicals needed by living things. Animals, almost by definition, obtain these other chemicals along with the carbohydrates and proteins they ingest when they consume other animals or plants, therefore, animals are usually content as long as the water has enough O_2 and a decent salinity. Plants, on the other hand, are more self-sufficient, and they can synthesize a wide variety of complex molecules from simple inorganic precursors. The ecological community that will develop in a body of water is thus often dependent on the suitability of the habitat for the growth of photosynthetic organisms. Exceptions to this include the deep ocean (which is dependent, however, on the growth of phytoplankton above), or headwater streams (which depend on adjacent trees for most of their organic input through leaffall), or cave streams (bat guano) and so on.

Along with sunlight and CO_2, the major needs of a plant include macronutrients nitrogen and phosphorous (used for proteins, DNA, RNA, ATP, etc.) and micronutrients such as sulfur (protein), magnesium (chlorophyll), and iron (cytochromes) (this list is not comprehensive). The micronutrients may be found in very small concentrations; plants are good at obtaining them even if they are in low concentration; and they are rarely a limiting factor. Some micronutrients, such as sulfur (as SO_4^{-2}), magnesium (as Mg^{+2}), calcium (as Ca^{+2}) and potassium (as K^+) are important constituents of both seawater and freshwater.

Phosphorous (P) and nitrogen (N) are critical to plant growth, and they (usually P, but sometimes N) are often limiting factors to plant growth. Before you object, recalling that dissolved nitrogen is common in water, remember that it is dissolved nitrogen *gas*, N_2, which is inert and cannot be used by most plants. The exception here are the cyanobacteria, which can fix N_2 in the heterocysts, which provide a local anoxic environment for the nitrogen-fixing enzymes (and bacteria in anoxic root nodules of legumes and other anoxic places in the soil). Bodies of water with a low N/P ratio are thus prone to blooms of cyanobacteria. For most plants, N must be in the form of nitrate (NO_3^-) or ammonia (NH_3, NH_4^+ in water). Ammonia, of course, is the nitrogenous waste of choice for many aquatic organisms, and even more is released by bacteria breaking down dead plants and animals or other nitrogenous animal wastes such as urea. Nitrate is a product of the nitrogen cycle; the nitrogen cycle in water differs slightly from the nitrogen cycle that takes place on land (which you are probably familiar with).

On land, N_2 is fixed by bacteria in the soil such as *Rhizobium, Clostridium,* and *Azotobacter*; in the water (both freshwater and marine) N_2 is fixed by such cyanobacteria as *Anabaena, Plectonema,* and *Nostoc*. The reaction requires energy and proceeds as follows:

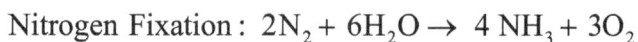

$$\text{Nitrogen Fixation}: 2N_2 + 6H_2O \rightarrow 4NH_3 + 3O_2$$

Ammonia, whether generated by nitrogen fixation or by the breakdown of amino acids by animals or decomposers, is toxic. As the pH of water increases, more of the ammonia exists in the water as NH_4^+. NH_4^+ is even more toxic than NH_3, and the fact that it is more prevalent at higher pH leads to one of the significant differences between keeping a marine and a freshwater aquarium. A marine aquarium typically has a pH of 8.0 to 8.5; a freshwater aquarium will usually have a pH of about 7. At pH 8.0, there is far more NH_4^+ present, and, if too many animals are producing too much NH_3, then NH_4^+ levels will soon become toxic. Therefore, marine tanks must be "aged", that is, stocked slowly, to allow populations of bacteria to develop to remove the ammonia. The role of these latter bacteria will be explored below.

Once produced, ammonia (NH_3) is used by a variety of plants and bacteria as the source of the amino group for amino acid synthesis (another reason that freshwater tanks are more tolerant than marine tanks in regards to ammonia is the ready availability of freshwater plants which help reduce ammonia levels). The amino acid synthesis reaction also requires energy and looks like this:

$$\text{Amino acid synthesis: } 2NH_3 + 2H_2O + 4CO_2 \rightarrow 2\ CH_2NH_2COOH + 3O_2$$

Note that both this reaction and the preceding one release O_2 into the atmosphere; photosynthesis is not the only source of O_2 in the atmosphere! While amino acid synthesis does remove some ammonia from the water, much more is usually present. Another reaction, *nitrification*, takes ammonia and converts it to *nitrite* (NO_2^-); this reaction releases energy to the organism which carries it out:

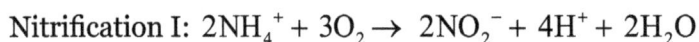

$$\text{Nitrification I: } 2NH_4^+ + 3O_2 \rightarrow\ 2NO_2^- + 4H^+ + 2H_2O$$

In water, this reaction is carried out mostly by bacteria of the genus *Nitrosomonas*. Nitrite is less toxic than ammonia, but is still toxic; high levels of nitrite can kill many aquatic organisms. Fortunately, a further nitrification reaction can occur (also with a release of energy):

$$\text{Nitrification II: } 2NO_2^- + O_2 \rightarrow\ 2NO_3^-$$

The end product here, *nitrate* (NO_3^-), is even less toxic than nitrite, and can be used by many plants as a nitrogen source. In aquatic systems and terrestrial systems as well, this reaction is carried out by bacteria of the genus *Nitrobacter*. In a typical marine aquarium, nitrate may approach toxic levels, but this process takes months. In addition, a number of *denitrification* reactions take place and reduce nitrate levels, as does uptake by plants.

To retrace the nitrogen cycle, let us consider the marine aquarium again. Ammonia levels build as animals excrete nitrogenous wastes; as they die and decompose; as food (with protein) is added; and as N_2 from the atmosphere is fixed by cyanobacteria. Because of the high pH (8.0), most of the ammonia will exist as toxic NH_4^+. *Nitrosomonas* bacteria will convert the ammonia to nitrite, and *Nitrobacter* bacteria will convert the nitrite to nitrate, which can be utilized by plants. Denitrification will remove some of the nitrate from the water. In a freshly established marine tank with a few fish, it is not uncommon for the ammonia levels to peak, then drop as the *Nitrosomonas* bacteria take hold and begin to convert ammonia to nitrite. As nitrite levels build and peak, *Nitrobacter* populations will thrive and convert the nitrite to nitrate, reducing nitrite concentrations to near zero. Typically, it takes about one month for the bacterial populations to become established, and it is usually wise to monitor the process by daily tests of ammonia and nitrite levels. The number of organisms that can be maintained in a marine tank is usually proportional to the amount of surface area on the gravel of the aquarium available for the *Nitrosomonas* and *Nitrobacter* bacteria to attach to.

Diagram of the Nitrogen Cycle in water. N_2 in the air is fixed by cyanobacteria and put in the form of protein, which is eaten by fish. Fish release NH_3, which is taken up by heterotrophic bacteria, plants, cyanobacteria, and the bacterium *Nitrosomonas*. The first three reincorporate the NH_3 into protein, *Nitrosomonas* converts it to NO_2, which is taken up in turn by *Nitrobacter*, which gives off NO_3. The NO_3 is a plant nutrient and is also utilized by anaerobic bacteria, which can produce N_2, NO_2, or NH_3.

The Nitrogen Cycle in Water

NH_3 Animals Plants Cyanobacteria N_2

(Proteins)

Heterotrophic Bacteria Nitrosomonas NO_2

Nitrobacter

Anaerobic Bacteria NO_3

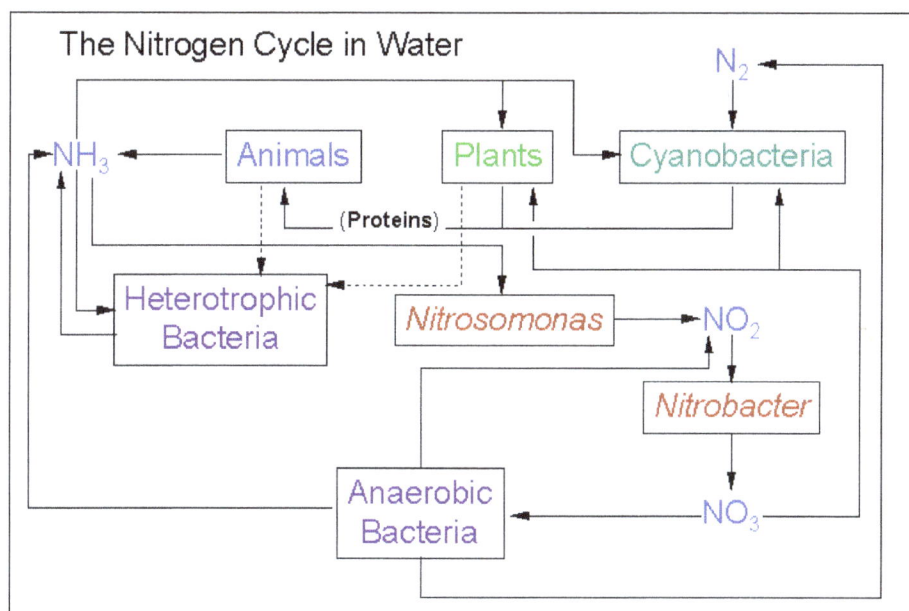

The story with phosphorous is much simpler. Phosphorus is typically available to plants as *organophosphate* (PO_4^{-3}); this compound is a common weathering product of igneous rock. Other sources of phosphate are decaying animal bodies, animal wastes, bones of vertebrates, and bird guano. The last is mentioned specifically because it is such a rich source of phosphorous; in fact, huge quantities of bird guano that accumulate near nesting areas of shorebirds are often mined for fertilizer. Phosphorous trapped in the bodies of dead organisms which sink to the bottom may accumulate in bottom sediments; this phosphorous will become available again to living plants (animals do not take up phosphorous directly from the environment) when currents sweep the bottom and bring it to the surface where there is sufficient light for photosynthesis. In nature, two such currents are noteworthy, the *overturn* of lakes, and *upwellings*, places where cold ocean currents meet continents and rise up from the bottom. The most famous example of an upwelling is off the coast of Peru; as the cold water reaches the surface (replacing warm surface waters blown west by prevailing winds), it brings up phosphorous, which encourages the growth of abundant algae, which in turn are appetizers for anchovies, which are then fished in great numbers by humans (for pizza topping) and birds.

Phosphorous is often the limiting factor for plant growth in aquatic systems. In many aquatic habitats, it is impossible to measure any free phosphate in the water when heavy algal growth (blooms) are in progress, mostly because the algae use up the phosphate as soon as it becomes available. Many algae store inclusions of phosphorous (when available) in their cells as a "hedge" against later phosphate shortages. Tremendous algal blooms are a symptom of *eutrophication*, a natural process that occurs as lakes age and accumulate phosphorous. Such huge blooms of algae can cause problems when they die and decompose, or when, particularly on hot, still summer nights, their respiration uses up all the O_2 in the water.

Many human activities accelerate the process of eutrophication; this leads to a phenomenon known as *cultural eutrophication* to set it apart from the natural process. In cultural eutrophication, phosphorous from agricultural runoff (phosphate fertilizers, animal wastes), human sewage, detergents, etc., is added to the water. The much publicized "death" of the Great Lakes was (is) due to cultural eutrophication. A eutrophied lake is typified by heavy algal growth, turbid water, and overall high productivity. The opposite, an *oligotrophic* lake, has clear water with little life (paradoxically, highly eutrophic lakes such as our "dead" Great Lakes, are teeming with life, while pristine oligotrophic lakes are nearly sterile). Remedies to cultural eutrophication include soil erosion control (much phosphorus is carried into aquatic systems bound to soil particles), careful use of fertilizers (the use of fertilizers on grass is criminal and should be totally banned), and *tertiary* sewage treatment (primary treatment kills germs, secondary removes solids, tertiary removes phosphorus and other chemicals). Even after the source of phosphorus is found and diverted, the sediments may contain enough phosphorus to keep the lake eutrophic for years. Again, under natural conditions, oligotrophic lakes occur near headwaters where little weathering has occurred to release phosphorous, and eutrophic lakes are found downstream (downstream is usually geologically older), where phosphorous inputs from the upstream sites make abundant algal growth possible.

Biosphere

Biosphere is the layer of the planet Earth where life exists. This layer ranges from heights of up to ten kilometres above sea level, used by some birds in flight, to depths of the ocean such as the Puerto Rico trench, at more than 8 kilometres deep. These are the extremes; however, in general the layer of the Earth containing life is thin: the upper atmosphere has little oxygen and very low temperatures, while ocean depths greater than 1000 m are dark and cold. In fact, it has been said that the biosphere is like the peel in relation to the size of an apple.

The biosphere is unique. So far there has been no existence of life elsewhere in the universe. Life on Earth depends on the sun. Energy, provided as sun light, is captured by plants, some bacteria and protists, in the marvellous phenomenon of photosynthesis. The captured energy transforms carbon dioxide into organic compounds such as sugars and produces oxygen. The vast majority of species of animals, fungi, parasitic plants and many bacteria depend directly or indirectly on photosynthesis.

The Diversity of Life

The biosphere supports between 3 and 30 million species of plants, animals, fungi, single-celled prokaryotes such as bacteria, and single-celled eukaryotes such as protozoans. Of this total, only about 1.4 million species have been named so far, and fewer than 1 percent have been studied for their ecological relationships and their role in ecosystems. A little more than half the named species are insects, which dominate terrestrial and freshwater communities worldwide; the laboratories of systematists are filled with insect species yet to be named and described. Hence, the relationships of organisms to their environments and the roles that species play in the biosphere are only beginning to be understood.

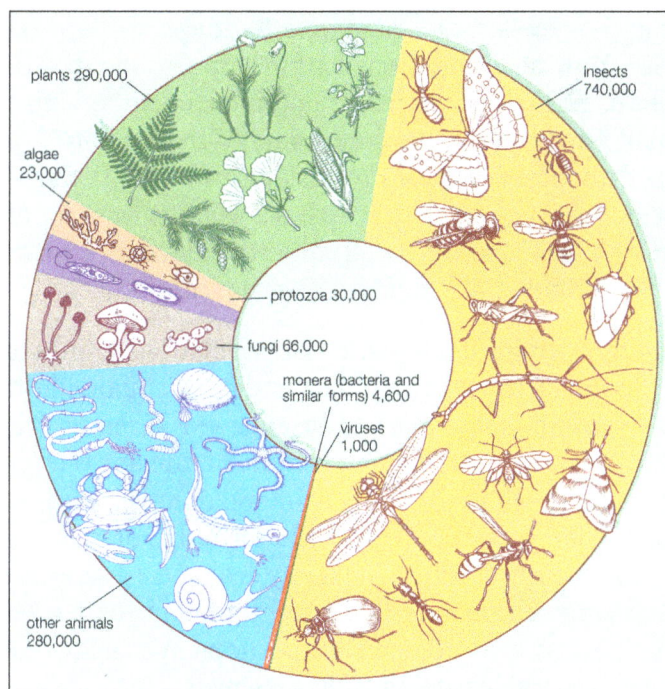

Estimated number of known living species. The majority of species
are still unknown—i.e., yet to be described by taxonomists.

The Organization of the Biosphere

Natural Groupings

This tremendous diversity of life is organized into natural ecological groupings. As life has evolved, populations of organisms have become separated into different species that are reproductively isolated from one another. These species are organized through their interrelationships into complex biological communities. The interactions in these communities affect, and are affected by, the physical environments in which they occur, thereby forming ecosystems through which the energy and nutrients necessary for life flow and cycle. The mix of species and physical environments vary across the globe, creating ecological communities, or biomes, such as the boreal forests of North America and Eurasia and the rainforests of the tropics. The sum total of the richness of these biomes is the biosphere.

Processes of Evolution

This hierarchical organization of life has come about through the major processes of evolution— natural selection (the differential success of the reproduction of hereditary variations resulting from the interaction of organisms with their environment), gene flow (the movement of genes among different populations of a species), and random genetic drift (the genetic change that occurs in small populations owing to chance). Natural selection operates on the expressed characteristics of genetic variants found within populations, winnowing members of the population who are less well suited to their environment from those better suited to it. In this manner, populations become adapted to their local ecosystems, which include both the physical environment and the other species with which they interact in order to survive and reproduce.

The genetic variation that is necessary for a species to adapt to the physical environment and to other organisms arises from new mutations within populations, the recombination of genes during sexual reproduction, and the migration of and interbreeding with individuals from other populations. In very small populations, however, some of that variation is lost by chance alone through random genetic drift. The combined result of these evolutionary processes is that after many generations populations of the same species have widely divergent characteristics. Some of these populations eventually become so genetically different that their members cannot successfully interbreed, resulting in the evolution of a separate species (speciation).

The diversification of life through local adaptation of populations and speciation has created the tremendous biodiversity found on the Earth. In most regions a square kilometre (0.4 square mile) will harbour hundreds—in some places even thousands—of species. The interactions between these species create intricate webs of relationships as the organisms reciprocally evolve, adapting to one another and becoming specialized for their interactions. Natural communities of species reflect the sum of these species' interactions and the ongoing complex selection pressures they constantly endure that drive their evolution. The many ecological and evolutionary processes that affect the relationships among species and their environments render ecology one of the most intricate of the sciences. The answers to the major questions in ecology require an understanding of the relative effects of many variables acting simultaneously.

The Importance of the Biosphere

The continued functioning of the biosphere is dependent not only on the maintenance of the intimate interactions among the myriad species within local communities but also on the looser yet crucial interactions of all species and communities around the globe. The Earth is blanketed with so many species and so many different kinds of biological communities because populations have been able to adapt to almost any kind of environment on Earth through natural selection. Life-forms have evolved that are able to survive in the ocean depths, the frigid conditions of Antarctica, and the near-boiling temperatures of geysers. The great richness of adaptations found among different populations and species of living organisms is the Earth's greatest resource. It is a richness that has evolved over millions of years and is irreplaceable.

It is therefore startling to realize that our inventory of the Earth's diversity is still so incomplete that the total number of living species cannot be estimated more closely than between 3 and 30 million species.

Despite the limited scientific knowledge of most species, ecological studies during the 20th century made great headway in unraveling the mechanisms by which organisms coevolve with one another and adapt to their physical environment, thereby shaping the biosphere. Each new decade has produced a steady stream of studies showing that the biological and physical elements of the Earth are more interconnected than had been previously thought. Those studies also have shown that often the most seemingly insignificant species are crucial to the stability of communities and ecosystems. Many seemingly obscure species are at risk worldwide of being dismissed as unimportant. The effect that the loss of species will have on ecosystems is appreciated only by understanding the relationships between organisms and their environments and by studying the ecological and evolutionary processes operating within ecosystems.

The need to understand how the biosphere functions has never been greater. When human population levels were low and technological abilities crude, societies' impact on the biosphere was relatively small. The increase in human population levels and the harvesting of more of the Earth's natural resources has altered this situation, especially in recent decades. Human activities are causing major alterations to the patterns of energy flow and nutrient cycling through ecosystems, and these activities are eliminating populations and species that have not even been described but which might have been of central importance to the maintenance of ecosystems.

The biologist Edward O. Wilson, who coined the term biodiversity, estimated conservatively that in the late 20th century at least 27,000 species were becoming extinct each year. The majority of these were small tropical organisms. The impact that this freshet of extinctions would have on the biosphere is akin to receiving a box of engine parts and discarding a portion of them before reading the directions, assuming that their absence will have no negative repercussions on the running of the engine.

Resources of the Biosphere

Photosynthetic Process

Life on Earth depends on the harnessing of solar energy by the process of photosynthesis. Photosynthetic plants convert solar energy into the chemical energy of living tissue, and that stored chemical energy flows into herbivores, predators, parasites, decomposers, and all other forms of life. In the photosynthetic process, light energy is absorbed by the chlorophyll molecules of plants to convert carbon dioxide and water into carbohydrates and oxygen gas. Proteins, fats, nucleic acids, and other compounds also are synthesized during the process, as long as elements such as nitrogen, sulfur, and phosphorus are available.

Efficiency of Solar Energy Utilization

Most solar energy occurs at wavelengths unsuitable for photosynthesis. Between 98 and 99 percent of solar energy reaching the Earth is reflected from leaves and other surfaces and absorbed by other molecules, which convert it to heat. Thus, only 1 to 2 percent is available to be captured by plants. The rate at which plants photosynthesize depends on the amount of light reaching the leaves, the temperature of the environment, and the availability of water and other nutrients such as nitrogen and phosphorus. The measurement of the rate at which organisms convert light energy (or inorganic chemical energy) to the chemical energy of organic compounds is called primary productivity. Hence, the total amount of energy assimilated by plants in an ecosystem during photosynthesis (gross primary productivity) varies among environments. (Productivity is often measured by an increase in biomass, a term used to refer to the weight of all living organisms in an area. Biomass is reported in grams or metric tons.)

Much of the energy assimilated by plants through photosynthesis is not stored as organic material but instead is used during cellular respiration. In this process organic compounds such as carbohydrates, proteins, and fats are broken down, or oxidized, to provide energy (in the form of adenosine triphosphate [ATP]) for the cell's metabolic needs. The energy not used in this process is stored in plant tissues for further use and is called net primary productivity. About 40 to 85 percent of gross primary productivity is not used during respiration and becomes net primary productivity.

The highest net primary productivity in terrestrial environments occurs in swamps and marshes and tropical rainforests; the lowest occurs in deserts. In aquatic environments, the highest net productivity occurs in estuaries, algal beds, and reefs. Consequently, these environments are especially critical for the maintenance of worldwide biological productivity.

Energy Transfers and Pyramids

A small amount of the energy stored in plants, between 5 and 25 percent, passes into herbivores (plant eaters) as they feed, and a similarly small percentage of the energy in herbivores then passes into carnivores (animal eaters). The result is a pyramid of energy, with most energy concentrated in the photosynthetic organisms at the bottom of food chains and less energy at each higher trophic level (a division based on the main nutritional source of the organism; see community ecology: Trophic pyramids and the flow of energy). Some of the remaining energy does not pass directly into the plant-herbivore-carnivore food chain but instead is diverted into the detritus food chain. Bacteria, fungi, scavengers, and carrion eaters that consume detritus (detritivores) are all eventually consumed by other organisms.

The rate at which these consumers convert the chemical energy of their food into their own biomass is called secondary productivity. The efficiency at which energy is transferred from one trophic level to another is called ecological efficiency. On average it is estimated that there is only a 10 percent transfer of energy.

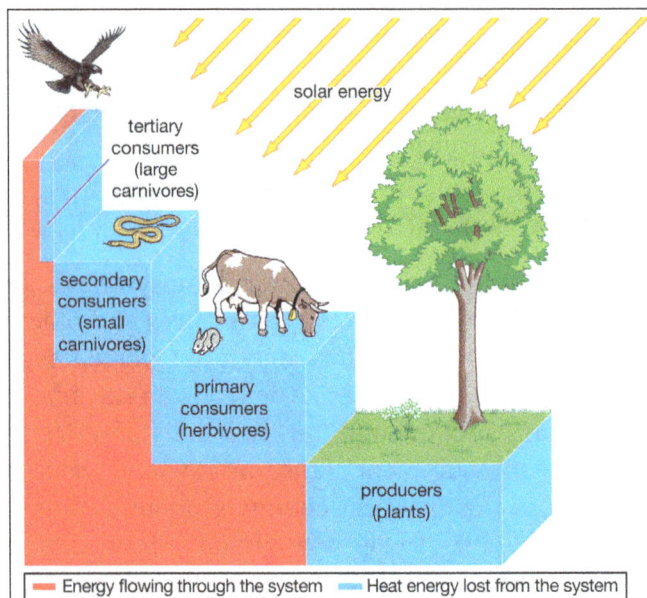

Transfer of energy through an ecosystem. At each trophic level only a small proportion of energy (approximately 10 percent) is transferred to the next level.

Energy is lost in several ways as it flows along these pathways of consumption. Most plant tissue is uneaten by herbivores, and this stored energy is therefore lost to the plant-herbivore-carnivore food chain. In terrestrial communities less than 10 percent of plant tissue is actually consumed by herbivores. The rest falls into the detritus pathway, although the detritivores consume only some of this decaying tissue. Oil and coal deposits are major repositories of this unused plant energy and have accumulated over long periods of geologic time.

The efficiency by which animals convert the food they ingest into energy for growth and reproduction is called assimilation efficiency. Herbivores assimilate between 15 and 80 percent of the plant material they ingest, depending on their physiology and the part of the plant that they eat. For example, herbivores that eat seeds and young vegetation high in energy have the highest assimilation efficiencies, those that eat older leaves have intermediate efficiencies, and those that feed on decaying wood have very low efficiencies. Carnivores generally have higher assimilation efficiencies than herbivores, often between 60 and 90 percent, because their food is more easily digested.

The overall productivity of the biosphere is therefore limited by the rate at which plants convert solar energy (about 1 percent) into chemical energy and the subsequent efficiencies at which other organisms at higher trophic levels convert that stored energy into their own biomass (approximately 10 percent). Human-induced changes in net primary productivity in the parts of the biosphere that have the highest productivity, such as estuaries and tropical moist forests, are likely to have large effects on the overall biological productivity of the Earth.

Environmental Conditions

Most organisms are limited to either a terrestrial or an aquatic environment. An organism's ability to tolerate local conditions within its environment further restricts its distribution. One parameter, such as temperature tolerance, may be important in determining the limits of distribution, but often a combination of variables, such as temperature tolerance and water requirements, is important. Extreme environmental variables can evoke physiological and behavioral responses from organisms. The physiological response helps the organism maintain a constant internal environment (homeostasis), while a behavioral response allows it to avoid the environmental challenge—a fallback strategy if homeostasis cannot be maintained.

The ways in which modern living organisms tolerate environmental conditions reflect the aquatic origins of life. With few exceptions, life cannot exist outside the temperature range at which water is a liquid. Thus, liquid water, and temperatures that maintain water as a liquid, are essential for sustaining life. Within those parameters, the concentrations of dissolved salts and other ions, the abundance of respiratory gases, atmospheric or hydrostatic pressure, and rate of water flow all influence the physiology, behaviour, and distribution of organisms.

Temperature

Temperature has the single most important influence on the distribution of organisms because it determines the physical state of water. Most organisms cannot live in conditions in which the temperature remains below 0 °C or above 45 °C for any length of time. Adaptations have enabled certain species to survive outside this range—thermophilic bacteria have been found in hot springs in which the temperatures may approach the boiling point, and certain polar mosses and lichens can tolerate temperatures of −70 °C—but these species are the exceptions. Few organisms can remain for long periods at temperatures above 45 °C, because organic molecules such as proteins will begin to denature. Nor are temperatures below freezing conducive to life: cells will rupture if the water they contain freezes.

Most organisms are not able to maintain a body temperature that is significantly different from that of the environment. Sessile organisms, such as plants and fungi, and very small organisms and animals that cannot move great distances, therefore, must be able to withstand the full range of temperatures sustained by their habitat. In contrast, many mobile animals employ behavioral mechanisms to avoid extreme conditions in the short term. Such behaviours vary from simply moving short distances out of the Sun or an icy wind to large-scale migrations.

Some types of animals employ physiological mechanisms to maintain a constant body temperature, and two categories are commonly distinguished: the term cold-blooded is understood to refer to reptiles and invertebrates, and warm-blooded is generally applied to mammals and birds. These terms, however, are imprecise; the more accurate terms, ectotherm for cold-blooded and endotherm for warm-blooded, are more useful in describing the thermal capabilities of these animals. Ectotherms rely on external sources of heat to regulate their body temperatures, and endotherms thermoregulate by generating heat internally.

Terrestrial ectotherms utilize the complex temperature profile of the terrestrial environment to derive warmth. They can absorb solar radiation, thus raising their body temperatures above that of the surrounding air and substrate, unlike the aquatic ectotherm, whose body temperature is usually very close to that of the environment. As this solar radiation is taken up, physiological mechanisms contribute to the regulation of heat—peripheral blood vessels dilate and heart rate increases. The animal also may employ behavioral mechanisms, such as reorienting itself toward the Sun or flattening its body and spreading its legs to maximize its surface area exposure. At night, loss of heat may be reduced by other behavioral and physiological mechanisms—the heart rate may slow, peripheral blood vessels may constrict, surface area may be minimized, and shelter may be sought.

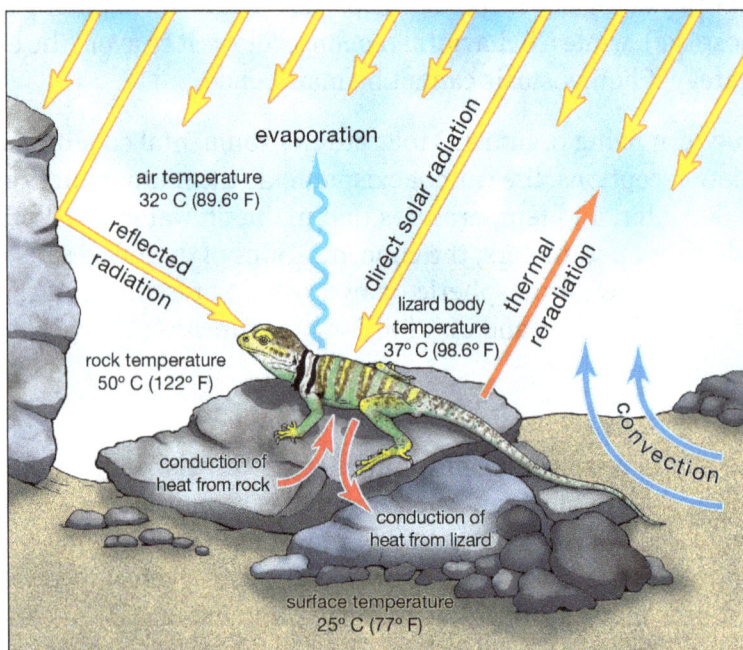

Energy exchange between a terrestrial reptile and the environment.

Endotherms maintain body temperature independently of the environment by the metabolic production of heat. They generate heat internally and control passive heat loss by varying the quality of their insulation or by repositioning themselves to alter their effective surface area (i.e., curling into a tight ball). If heat loss exceeds heat generation, metabolism increases to make up the loss.

If heat generation exceeds the rate of loss, mechanisms to increase heat loss by evaporation occur. In either case, behavioral mechanisms can be employed to seek a more suitable thermal environment.

To survive for a limited period in adverse conditions, endotherms may employ a combination of behavioral and physiological mechanisms. In cold weather, which requires an increase in energy consumption, the animal may enter a state of torpor in which its body temperature, metabolism, respiratory rate, and heart rate are depressed. Long-term winter hypothermia, or hibernation, is an extended state of torpor that some animals use as a response to cold conditions. Torpor and hibernation free the animals from energetically expensive maintenance of high body temperatures, saving energy when food is limited.

Another form of torpor, estivation, is experienced by animals in response to heat stress. This state is seen more often in ectothermic animals than in endotherms, but in both the stimulus for estivation is usually a combination of high temperatures and water shortage.

Humidity

Most terrestrial organisms must maintain their water content within fairly narrow limits. Water commonly is lost to the air through evaporation or, in plants, transpiration. Because most water loss occurs by diffusion and the rate of diffusion is determined by the gradient across the diffusion barrier such as the surface of a leaf or skin, the rate of water loss will depend on the relative humidity of the air. Relative humidity is the percent saturation of air relative to its total saturation possible at a given temperature. When air is totally saturated, relative humidity is said to be 100 percent. Cool air that is completely saturated contains less water vapour than completely saturated warm air because the water vapour capacity of warm air is greater. Diffusion gradients across skin or leaves, therefore, can be much steeper in summer when the air is warm, rendering evaporative water loss a much more serious problem in warm environments than in cool environments. Nevertheless, rates of water loss are higher in dry air (conditions of low relative humidity) than in moist air (conditions of high relative humidity), regardless of the temperature.

Water loss from evaporation must be compensated by water uptake from the environment. For most plants, transpirational water loss is countered by the uptake of water from the soil via roots. For animals, water content can be replenished by eating or drinking or by uptake through the integument. For organisms living in dry environments, there are many morphological and physiological mechanisms that reduce water loss. Desert plants, or xerophytes, typically have reduced leaf surface areas because leaves are the major sites of transpiration. Some xerophytes shed their leaves altogether in summer, and some are dormant during the dry season.

Desert animals typically have skin that is relatively impervious to water. The major site of evaporation is the respiratory exchange surface, which must be moist to allow the gaseous exchange of oxygen and carbon dioxide. A reduction in amount of water lost through respiration can occur if the temperature of the exhaled air is lower than the temperature of the body. As many animals, such as gazelles, inhale warm air, heat and water vapour from the nasal passages evaporate, cooling the nose and the blood within it. The cool venous blood passes close to and cools the warm arterial blood traveling to the brain. If the brain does not require cooling, the venous blood returns to the heart by another route. The nasal passages also cool the warm, saturated air from the lungs so that water condenses in the nose and is reabsorbed rather than lost to the environment.

pH

The relative acidity or alkalinity of a solution is reported by the pH scale, which is a measure of the concentration of hydrogen ions in solution. Neutral solutions have a pH of 7. A pH of less than 7 denotes acidity (an increased hydrogen ion concentration), and above 7 alkalinity (a decreased hydrogen ion concentration). Many important molecular processes within the cells of organisms occur within a very narrow range of pH. Thus, maintenance of internal pH by homeostatic mechanisms is vital for cells to function properly. Although pH may differ locally within an organism, most tissues are within one pH unit of neutral. Because aquatic organisms generally have somewhat permeable skins or respiratory exchange surfaces, external conditions can influence internal pH. These organisms may accomplish the extremely important task of regulating internal pH by exchanging hydrogen ions for other ions, such as sodium or bicarbonate, with the environment.

The pH of naturally occurring waters can range from very acidic conditions of about 3 in peat swamps to very alkaline conditions of about 9 in alkaline lakes. Naturally acidic water may result from the presence of organic acids, as is the case in a peat swamp, or from geologic conditions such as sulfur deposits associated with volcanic activity. Naturally occurring alkaline waters usually result from inorganic sources. Most organisms are unable to live in conditions of extreme alkalinity or acidity.

Salinity

The term salinity refers to the amount of dissolved salts that are present in water. Sodium and chloride are the predominant ions in seawater, and the concentrations of magnesium, calcium, and sulfate ions are also substantial. Naturally occurring waters vary in salinity from the almost pure water, devoid of salts, in snowmelt to the saturated solutions in salt lakes such as the Dead Sea. Salinity in the oceans is constant but is more variable along the coast where seawater is diluted with freshwater from runoff or from the emptying of rivers. This brackish water forms a barrier separating marine and freshwater organisms.

The cells of organisms also contain solutions of dissolved ions, but the range of salinity that occurs in tissues is more narrow than the range that occurs in nature. Although a minimum number of ions must be present in the cytoplasm for the cell to function properly, excessive concentrations of ions will impair cellular functioning. Organisms that live in aquatic environments and whose integument is permeable to water, therefore, must be able to contend with osmotic pressure. This pressure arises if two solutions of unequal solute concentration exist on either side of a semipermeable membrane such as the skin. Water from the solution with a lower solute concentration will cross the membrane diluting the more highly concentrated solution until both concentrations are equalized. If the salt concentration of an animal's body fluids is higher than that of the surrounding environment, the osmotic pressure will cause water to diffuse through the skin until the concentrations are equal unless some mechanism prevents this from happening.

Many marine invertebrates have the same osmotic pressure as seawater. When the salt concentration of their surroundings changes, however, they must be able to adjust. Two means of contending with this situation are employed, and, depending on how they regulate the salt concentrations of their tissues, organisms are classified as osmoregulators or osmoconformers. The osmotic concentration of the body fluids of an osmoconformer changes to match that of its external envionment, whereas an osmoregulator controls the osmotic concentration of its body fluids, keeping them constant in spite of external alterations. Aquatic organisms that can tolerate a wide range of external ion concentrations are called euryhaline; those that have a limited tolerance are called stenohaline.

Even if aquatic organisms have an integument that is relatively impermeable to water, as well as to small inorganic ions, their respiratory exchange surfaces are permeable. Hence, organisms occurring in water that has a lower solute concentration than their tissues (e.g., trout in mountain streams) will constantly lose ions to the environment as water flows into their tissues. In contrast, organisms in salty environments face a constant loss of water and an influx of ions.

Many mechanisms have evolved that deal with these problems. Because water cannot be readily pumped across cell membranes, salinity balance is usually maintained by actively transporting inorganic ions, usually sodium and chloride. This process consumes energy and can usurp a large portion of the energy budget of animals in very saline environments. In marine fish, gill cells pump ions out of the body into the sea, while in freshwater fish gill cells pump ions in the opposite direction. Passive water loss in marine fish is compensated primarily in one of two ways. Most bony fish drink copiously and excrete salt across the gills, while the majority of sharks artificially elevate the salt concentration of their tissues above that of seawater with urea and other organic molecules, allowing water to slowly and passively enter the body. Through their food and across their gills, freshwater fish replenish most of the ions they lose. They also produce large quantities of very dilute urine to excrete excess water that diffuses into their bodies.

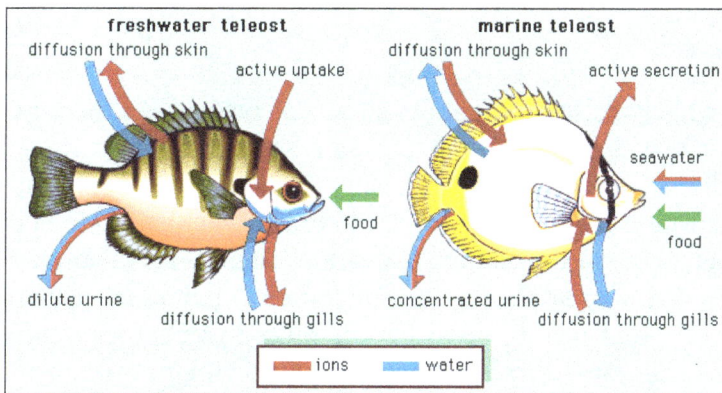

Osmotic regulation in freshwater and marine teleost fish.

Water Currents

The flow of water presents special problems for aquatic organisms. Flow is associated with rivers, oceanic currents, and waves and can be laminar (streamlined) or turbulent. Many organisms are specialized to live in flowing environments; the main obstacle to this lifestyle is the constant threat of being washed away. Both plants and animals have evolved mechanisms that help to anchor them to the substratum in flowing water (e.g., the holdfast of kelp or the byssus threads of mussels). If anchorage can be assured, there are many advantages to living in this environment. Flowing water generally is well oxygenated, and the supply is continuous; nutrients and food are constantly replenished as well. The very precariousness of the environment also affords some protection from predation because the number of predators that make this type of habitat home is limited.

Pressure

Atmospheric Pressure

Variations in atmospheric pressure can present special problems for the respiratory systems of animals because atmospheric pressure affects the exchange of oxygen and carbon dioxide that occurs during animal respiration. Normal atmospheric pressure at sea level is the total pressure that a column of air above the surface of the Earth exerts (760 millimetres of mercury, or 1 atmosphere). The total pressure is the sum of the pressures that each gas—mainly nitrogen, oxygen, and carbon dioxide—would exert alone (the partial pressure of that gas; see respiration: The gases in the environment). As an animal breathes, oxygen moves from the environment across the respiratory surfaces into the blood; carbon dioxide moves in the reverse direction. This process occurs primarily by passive diffusion; each gas moves from an area of greater to lesser partial pressure, driven by the differential that exists across the respiratory surface. At higher altitudes, where the atmospheric pressure is lower, the partial pressure of oxygen is also lower. The partial pressure differential of oxygen, therefore, is also lower, and the organism effectively receives less oxygen when it breathes, even though the percentage of oxygen in the air remains constant. This lack of oxygen is why humans carry oxygen when ascending to high altitudes. Humans who live in mountainous regions, however, can become acclimatized to the lowered availability of oxygen, and certain animals such as llamas have adaptations of the blood that allow them to live at high altitudes. Birds have very efficient lungs, and many apparently have no problems flying to high altitudes, even for extended flights.

Hydrostatic Pressure

Because air and water have vastly different densities, the pressures experienced in terrestrial and aquatic habitats differ markedly. A column of water, so much denser than air, exerts a greater amount of pressure than a column of air. With each 10-metre (32.8-foot) increase in depth, there is an increase in hydrostatic pressure equivalent to one atmosphere. Mean ocean depth is about 3,800 metres and has a pressure of about 380 atmospheres. To surmount this environmental challenge, animals that live at great depths lack air compartments such as lungs or swim bladders. Surface-dwelling animals that dive to great depths meet this challenge differently. As pressure increases during a dive, air compartments compress, returning to their former volume when the animal surfaces. Air is forced into the trachea, bronchi, and bronchioles, where no gas uptake occurs. Thus, the increased pressure cannot drive more gases into the bloodstream, and, as the animal rises, it does not experience the "bends" (decompression sickness resulting from a rapid reduction of air pressure). In contrast, sea snakes avoid the bends by excreting nitrogen across the skin to offset the uptake of this gas from the lungs.

Lithosphere

Lithosphere is the outermost layer of Earth consisting of crust and upper mantle portion of the earth.

Lithosphere is responsible for the hard and rigid surface of the Earth on which humans exists.

Lithosphere is not a continuous layer it is divided into the movable tectonic plates. This is 100 km deep layer.

The deepest and hottest part of the lithosphere is known as the asthenosphere.

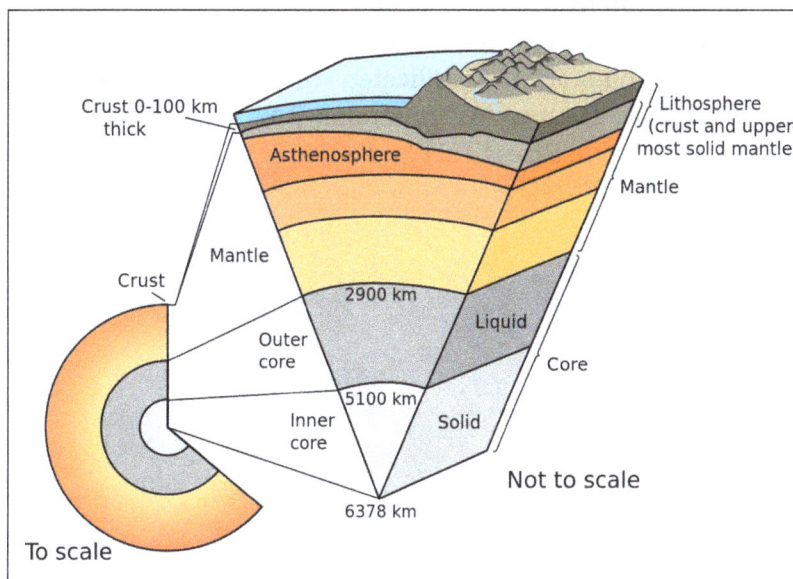

Oceanic Lithosphere

The oceanic lithosphere is composed primarily of mantle peridotites and magmatic rocks that are the result of the melting of these mantle rocks: gabbro, diabase and basalt.

Other rock types are present but in smaller quantities, such as secondary alteration products of these rocks (serpentinites), or hydrothermal deposits. The oceanic lithosphere is structured and emplaced at the axis of mid-ocean ridges, where magmatic, tectonic and hydrothermal processes operate and interact in response to plate separation. With ongoing plate separation the lithosphere is moved away from the narrow active axial zone. In some instances it can be modified by off-axis processes such as off-axis volcanism (hot spots), and recycled into the mantle after subduction. This simple tectonic history and limited number of rock types present in it make the structure and composition of the oceanic lithosphere a simple object to study when compared to that of continents. Most of the knowledge acquired on lithospheric processes is obtained from combined geological and geophysical observations, primarily seismic imaging.

One key aspect to understand the dynamics of mid-ocean ridges that shape the oceanic lithosphere is the rheology. The mechanical properties of the rocks and their mode of deformation control the nature of faulting and seismicity, shapes the porosity structure of the crust that allows water to penetrate and establish hydrothermal systems, or determines the mode of melt propagation in the mantle and crust. Lithospheric rheology is also strongly affected by the presence of fluids (water and other volatiles, and melt). The distribution and quantification of these fluid phases is poorly constrained by seismic methods, but electromagnetic studies, but additional electromagnetic studies are required as they are still scarce and only available at a limited number of sites.

Mid-ocean Ridges and the Oceanic Lithosphere

Map of seafloor spreading rate (colors) and seafloor age (contours), showing the main tectonic plates. The oceanic lithosphere (white and colored areas) covers >65% of the Earth's surface. Digital seafloor data compilation from Müller et al. Ridges are often classified according to their spreading rate, as marked. Plate boundaries are marked in red, and correspond to zero age crust along ridges; 10 Myr age contours are also indicated (thin blue lines). Plates: EU: Eurasia; SA: South America; AF: Africa; PA: Pacific; AU: Australia; AN: Antarctica; CO: Cocos; NZ: Nazca; NA:North America. The East Pacific Rise (EPR) Mid-Atlantic Ridge (MAR) and SouthWest Indian Ridge (SWIR) are also indicated for reference.

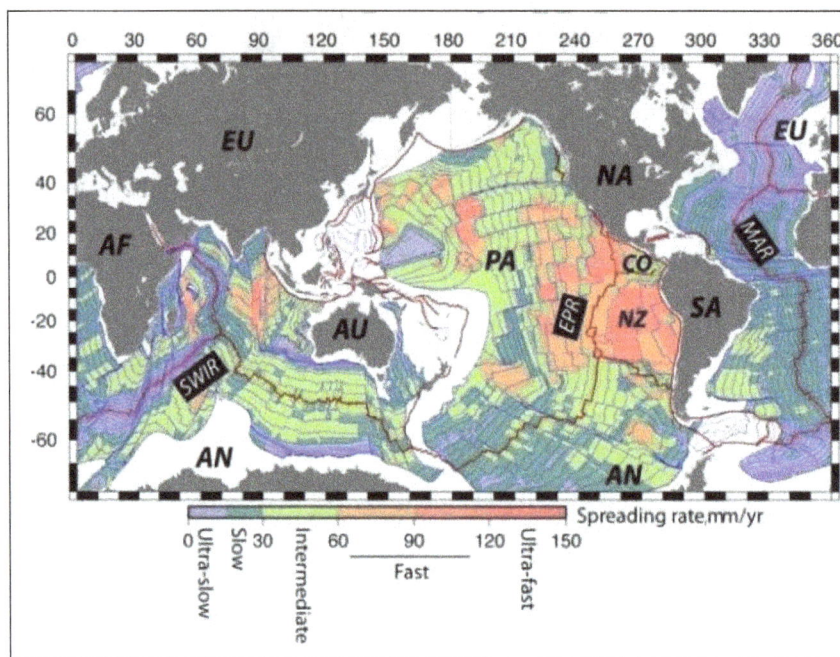

The oceanic lithosphere, which accounts for more than 65% of the Earth's surface, is formed along a narrow band of seafloor at the summit of mid-ocean ridges. Owing to its mode of formation and relatively simple history after formation, it differs substantially from the continental lithosphere. The mid-ocean ridges correspond to divergent plate boundaries, which have a total length of more than 60000 km and an elevation of 2 to 3 km above adjacent ocean basins. They are a consequence of mantle convection and upwelling driven by plate separation, and are initiated after rifting and break-up of continents. Most of the world's magmatism and seismicity concentrates along the axis of ridges, where a complex interaction between magmatic, tectonic, hydrothermal and biological processes takes place. The rate of plate separation is one of the main parameters that determine the architecture and composition of the oceanic lithosphere and associated active processes, and varies between near-zero cm/yr at the American-European plate boundary close the North pole, and >10 cm/yr at the East Pacific Rise between the Nazca and Pacific plates.

The separation of plates induces passive mantle flow under ocean basins and the ridge axis, and the resulting mantle upwelling controls the depth of ocean basins. Hot mantle material brought up under the axis is cooled from the top, and the isotherms deepen with age as the lithosphere moves away from the axis. The overlying seafloor bathymetry reflects this thermal structure by a gradual deepening away from the axis at the summit of mid ocean ridges. For seafloor ages younger than 70 Myr, the average global seafloor depth d is a function of time t:

$$d = 2.5 + 0.35t^{1/2}$$

with d in km and t in Myr. For seafloor ages older than 70 Myr the mean seafloor depth increases more slowly:

$$d = 6.4 - 3.2e^{-t/62.8}$$

Slightly different depth-age relationships are observed at different sites and ocean basins, reflecting variations in the temperature of the underlying mantle, or different thermal histories such as the presence or absence of hot spots. Further complications are introduced by lateral shifts of the position of the ridge axis, which are common throughout the history of ocean basins.

A second consequence of plate separation is pressure-release melting of the mantle during its ascent under the ridge axis. Mantle cools at a rate of 1-2° per ~3 km (1 kbar) of adiabatic ascent and crosses the mantle solidus at a depth that depends on the mantle temperature. As mantle continues to upwell, it melts ~1-2% per ~3 km of ascent. The total amount of melt produced under the ridge axis depends then on the thermal structure beneath the axis, that determines the shape of the melting regime and the thickness of the melting zone; larger amounts of melt are produced from a hotter mantle. Melt, which is lighter than the residual mantle, is collected and focused towards the ridge axis. The mode of emplacement of the melt depends on the shallow thermal structure of the oceanic lithosphere; at fast-spreading ridges all melt appears to gather and form a continuous, layered oceanic crust, while at slow spreading ridges the melt is emplaced in a discontinuous manner throughout the thickness of the lithosphere (T<750°C). This dependence of melt production on the thermal structure is most noticeable along 15-20% of the total length of the mid-ocean ridge system that is thermally influenced by hot spots (i.e., Galapagos or Iceland). Hot spots, which are generally considered zones of active mantle upwelling due to thermal anomalies in the mantle that are relatively fixed, also result in off-axis volcanism and formation of large oceanic plateaus and volcanic islands.

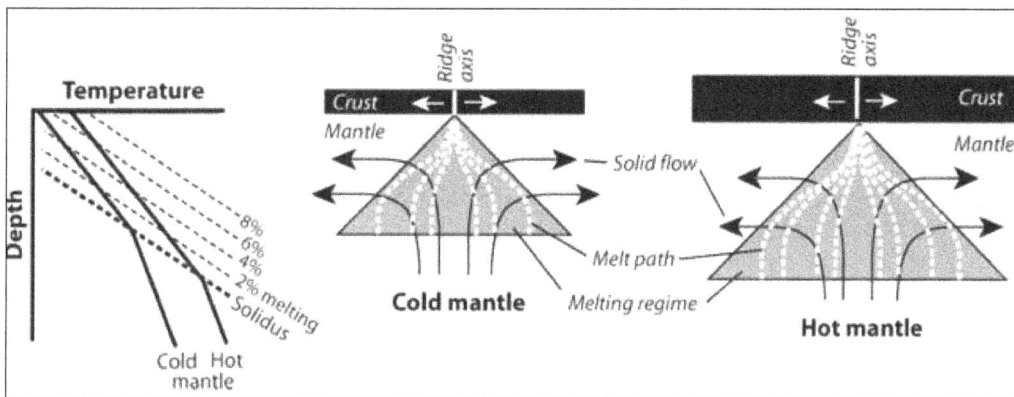

(Left) Mantle temperature decreases during adiabatic ascent under mid-ocean ridges. When the mantle temperature intersects the melting temperature (solidus, which is a function of pressure), melting initiates. Hotter mantle intersects the solidus at deeper levels, and undergoes more melting, than colder mantle. (Right) The melting regime is a function of the thermal structure of the lithosphere under the axis, and can be assimilated to a tent-shaped area in vertical cross-section. Lighter melt ascends and focuses to the ridge axis, decoupled from the passive solid flow. The collected forms a thicker crust in the case of a ridge with over hotter mantle (i.e., near hot spots). White arrows indicate the plate separation that drives passive mantle flow and upwelling under the axis.

The oceanic lithosphere is defined the outer shell of the Earth, that behaves as mechanical layer limited in depth by the brittle-to-plastic transition ($T \sim 750°C$). The crust has a compositional connotation, and in oceanic basins it is assimilated to a layered succession of gabbro, diabase and basalt, resulting from the mantle melting, which is emplaced above the Moho (mantle-crust transition). Discussion and description of the oceanic lithospheric structure also requires us to differentiate between the axial lithospheric structure, where magmatism and other processes are actively constructing and shaping the upper lithosphere, and the resulting off-axis, inactive lithosphere. The structure and composition of different types of common oceanic lithosphere away from the ridge axis, as inferred from both geological and geophysical observations. As active processes operate at the axis, the structure and composition of the lithosphere in these areas is substantially different from that of mature lithosphere, the characteristics of both magmatic and amagmatic ridges are discussed separately. Recent geological data demonstrate that in some environments, particularly along slow-spreading segments, there is no continuous crust, and instead a heterogeneous mixture of mantle (peridotite) and magmatic rocks are exposed instead. As the crust cannot be defined as such in some geological environments, we will refer to the composition and structure of the oceanic lithosphere, as it encompasses both typical 'oceanic crust', heterogeneous lithosphere, zones of transition from continental to oceanic lithosphere, and the hot spot influenced areas.

Methods of Study of the Oceanic Lithosphere

Geological observations: Geological observations both at the seafloor and at exposed pieces of oceanic lithosphere tectonically emplaced in continents (ophiolites) provide information on composition, structure and evolution. Seafloor observations include dredging, drilling, and ROV and submersible sampling. Seafloor imagery (video and photography) has become a powerful tool to map seafloor features and tectonic structures.

Petrology and geochemistry: Mineral assemblages and chemical compositions of basalts, peridotites, gabbros and other oceanic rocks provide clues on the processes that operate during and after the formation of the mantle lithosphere. These studies provide constraints on mantle composition and temperature, melting, magma accumulation and crystallization in the lithosphere, eruption of basalts, fluid circulation and water-rock interaction, chronology of volcanic events, among other processes.

Bathymetry and seafloor structure: Bathymetry and acoustic images (sonar) provide information on the tectonic structure, distribution of lava flows and other geological features, thermal subsidence, etc. Ship-mounted systems allow for surveys of large areas of the seafloor, but near-bottom instruments mounted on deep-tow systems, remotely operated vehicles (ROVs) and autonomous underwater vehicles (AUVs) provide highresolution images to study in detail volcanic and tectonic processes at the ridge axis. Global bathymetric information is now also available from combined global shipboard bathymetry and satellite-derived sea surface gravity.

Gravity and magneticstudies: Gravimetric and magnetic measurements provide information on the density and magnetic structure of the oceanic lithosphere. Potential data requires correction for navigation, vehicle and topography effects to obtain anomalies that are then inverted to obtain information on the internal structure of the lithosphere. Magnetic anomalies are particularly useful to determine the seafloor age and the history of plate movements. Satellite altimetry data is also used to obtain global sea-surface gravity maps. Near-bottom studies with AUVs and ROVs provide highresolution information on the detailed structure of the upper lithosphere.

Electromagnetic studies: Both passive and active-source electromagnetic experiments provide images of the conductivity structure of the oceanic lithosphere. These methods are particularly important to map the presence of electrically conductive fluids such, as magma or water from hydrothermal fluids below the ridge axis.

Seismic methods: The most used methods to infer the structure and composition of the oceanic lithosphere are seismic reflection and refraction. Ocean bottom seismometers and active sources are commonly used to infer the velocity (P- and S-wave) structure of the oceanic lithosphere that can be interpreted in terms of lithology. Active sources are normally deployed at the sea surface, but sources at the seafloor allow for highresolution seismic studies. Seismic reflection data provide higher-resolution images of the reflectivity structure of the lithosphere, and have been particularly useful to map the thickness of extrusives, the presence of magma chambers, or the presence or absence of Moho reflections. Other parameters such as the Poisson's ratio, the attenuation, or the anisotropy provide also information on the structure and processes. A more recent technique, seafloor compliance, can be used to infer the S-wave structure of the oceanic lithosphere. Ocean bottom seismometers and underwater hydrophones can also be used to monitor seismicity, and study both magmatic and tectonic activity at mid-ocean ridges.

Permeability: The permeability of the lithosphere is a critical component of fluid circulation and cooling with age. ODP drill holes are used to study the permeability structure of the lithosphere near mid-ocean ridges by sealing sections of drilled wells, monitoring of pressure changes, and pumping water to study pressure variations in adjacent wells.

Rock mechanics and rheology: The mechanical properties of the oceanic lithosphere (rheology) are determined based on geophysical observations and rock deformation experiments at elevated pressures and temperatures. Results of laboratory experiments are extrapolated to geological scales and conditions (pressure, temperature, strain rate, etc.) to construct models of strength and mechanical behavior of the lithosphere. Predictions from these rheological modes are compared to and tested with observations (i.e., distribution of seismicity, textures of naturally deformed rocks, tectonic structures, etc.).

Laboratory measurements of physical properties: Interpretation of geophysical data (gravity, magnetic, seismic, etc.) requires laboratory measurements of different rock types under natural temperature and pressure conditions. Measuring physical parameters (seismic velocity, density, electrical properties, permeability, etc.) is required to properly interpret the geophysical images and discriminate among rock types. As some rock types have very similar properties (i.e., the seismic velocity of gabbro and partially altered peridotite), determination of several physical properties of the lithosphere may be required to properly constrain its composition (e.g., combined seismic and electromagnetic studies).

Heat flow and hydrothermal studies: Heat loss is an important factor on the evolution and thermal structure of the oceanic lithosphere and underlying asthenosphere. Conductive heat loss can be measured with probes equipped with thermistors in areas where sediments blanket the seafloor. These probes can measure both the temperature distribution with depth (temperature gradients) and the thermal conductivity of the sediments, from which the heat flow can be calculated. These measurements provide information on convection cells in the sedimentary cover of the oceanic lithosphere, in addition to the Earth's heat loss. A large proportion of the heat is lost through fluid flow through the crust at the ridge axis, through hydrothermal vents and diffuse flow. Mapping of water signals associated with this hydrothermal activity and temperature measurements of vent temperatures provide information on the distribution and dynamics of such systems, but provide no accurate quantification of the heat loss associated with hydrothermal venting.

Study of active processes: Mid ocean ridges are the center of active magmatic, tectonic, hydrothermal and biological processes that interact. To understand this complex system, an integrated approach has been adopted by the scientific community in recent years to study selected portions of the mid-ocean ridge system to concentrate all types of studies (seismicity, geodesy, mapping and bathymetry, microbiology, ecology, etc.) over extended periods of time (>10 years). These studies include the deployment of seafloor instruments, and, in some cases, lay out of cables or buoys to transmit real-time data from the seafloor.

Continental Lithosphere

Continental lithosphere is composed approximately of a 20- to 35-mile layer of light granitic crust underlain by a 60- to 80-mile layer of heavy mantle peridotite (an olivine-, pyroxene-rich rock). Recall that these layers together form a tectonic plate. Lithospheric delamination refers to any process whereby the mantle layer peels away from the crustal layer and sinks, on its own, into the mantle. The process occurs where the mantle lithosphere is heavier than the deep mantle rocks below it. The delaminated layer is replaced by an upwelling of warm mantle rock, which, in turn, warms the overlying crust resulting in thermal and isostatic uplift and possibly crustal extension and gravitational collapse. Delamination may have produced normal faults in the southern Appalachians just prior to collision of North America and Gondwana (Africa) during Alleghany orogeny as shown in Figure.

A form of lithospheric delamination known as lithospheric drip is a process by which the heavy lower lithosphere becomes unstable and sinks into the mantle in the shape of teardrops as depicted in Figure. This type of process is believed to have occurred below the Sierra Nevada between 11 and 3.5 million years ago, allowing hot asthenosphere to rise, causing volcanism and possibly as much as than 3,000 feet of surface uplift. In this case, the addition of heat coupled with the loss of a high-density lithospheric root resulted in a pulse of uplift. However, at about the same time, extension (normal faulting) occurred in Owens Valley to the immediate east of the Sierra Nevada, and compression occurred to the west in the Coast Ranges. From this example alone, we can clearly understand that the processes involved in mountain building and collapse are complex across an orogenic belt and that it is the interplay among heat, isostasy, plate movement, erosion, and mantle dynamics that governs how the Earth's surface will respond.

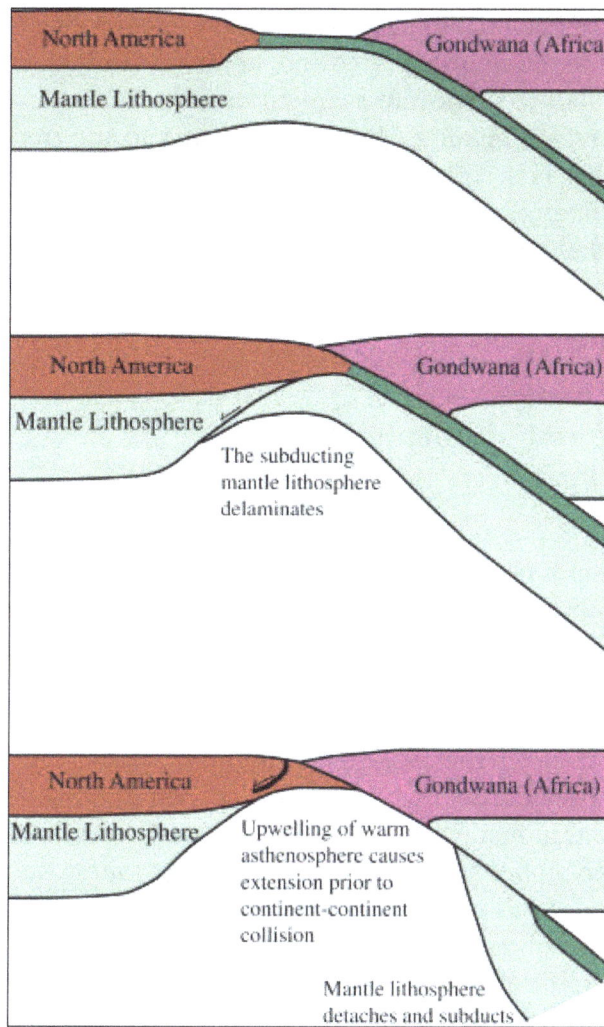

A series of cross-sections that show lithospheric delamination
and normal faulting just prior to the Allegheny orogeny.

Continental Subduction

Continental lithosphere can reach the subduction zone both in collisional processes, after the closure of an oceanic basin, and as patches of overthickened oceanic crust isostatically more buoyant than normal oceanic lithosphere (i.e., seamount chains, island arcs, plateaus). The former case have been approached by Regard et al. with a setup made of sand and silicone putty for the brittle and the ductile part of the lithosphere, respectively and honey for the mantle. Different sand'silicone layers, lighter or denser than the honey, simulated continental or oceanic lithospheres, respectively. Experimental outcomes showed that the oceanic closure is followed by subduction of continental lithosphere which may reach depths ranging between 50 and 450 km, prior to the onset of continental collision. This result is in agreement with the hydrocarbon on water Chemenda et al.'s models which showed how a large amount (200–300 km) of continental lithosphere can subduct if crust delaminates from the lithospheric mantle.

The amount of subductable continental material depends on the amount of subducted oceanic material, on the thickness of the convective mantle, and on the rheology of the slab.

If a buoyant segment of oceanic lithosphere reaches the trench, it can be subducted with the resulting effect of a decrease of the slab-pull force and, in turn, of the velocity of subduction. Models show, however, that the subducted buoyant segment must be large enough to alter the dynamics of the process. Hence, narrow oceanic ridges perpendicular to the trench are pulled down into subduction by the rest of the plate, without markedly modify the slab geometry as it happens above large subducted buoyant heterogeneities such as oceanic plateaus or ridges parallel to the trench. Experimental results highlighted also the unfeasibility to obtain a flat subduction only driven by buoyancy effects. Flat slab segments likely result both from the motion of the overriding plate towards the trench, and from the subduction of buoyant anomalies sufficiently large to perturb the local slab dynamics.

Thickness of Continental Lithosphere

The continental lithosphere varies considerably in thickness depending on its age and mechanism of formation. S-wave tomographic studies of the upper mantle have been most definitive in estimating the thickness of continental lithosphere. Most post-Archean lithosphere is 100 to 200 km thick, and lithosphere beneath Archean shields is commonly more than 300 km thick. Rheological models suggest thicknesses in these same ranges. In an S-wave tomographic cross-section around the globe, high-velocity roots underlie Archean crust, as in northern Canada, central and southern Africa, and Antarctica. The base of the lithosphere in these and other areas overlain by Archean crust may nearly reach the 410-km discontinuity. Under Proterozoic shields, however, lithospheric thicknesses rarely exceed 200 km. Consideration of elongation directions of Archean cratons relative to directions of modern plate motions suggests that the thick Archean lithosphere does not aid or hinder plate motions. As expected, hotspots (plumes), such as Hawaii and Iceland, are associated with slow velocities between 50 and 200 km deep. Thermal and geochemical modeling has shown that the lithosphere can be thinned by as much as 50 km by extension over mantle plumes. Isotopic and geochemical data from mantle xenoliths indicate that the mantle lithosphere beneath Archean shields formed during the Archean and that it is chemically distinct from post-Archean lithosphere. Because of the buoyant nature of the depleted Archean lithosphere, it tends to ride high compared with adjacent Proterozoic lithosphere, as shown by the extensive platform sediment cover on Proterozoic cratons compared with Archean cratons. The thick roots of Archean lithosphere often survive later tectonic events and thermal events, such as continental collisions and supercontinent rifting. However, mantle plumes or extensive later reactivation can remove the thick lithosphere keels, as for instance is the case with the Archean Wyoming province in North America and the north China craton.

Geosphere

The geosphere includes the rocks and minerals on Earth – from the molten rock and heavy metals in the deep interior of the planet to the sand on beaches and peaks of mountains. The geosphere also includes the abiotic (non-living) parts of soils, and the skeletons of animals that may become fossilized over geologic time.

Beyond these parts, the geosphere is about processes. The processes of the rock cycle such as metamorphism, melting and solidification, weathering, erosion, deposition, and burial are responsible for a constant recycling of rocks on Earth between sedimentary, igneous, and metamorphic states.

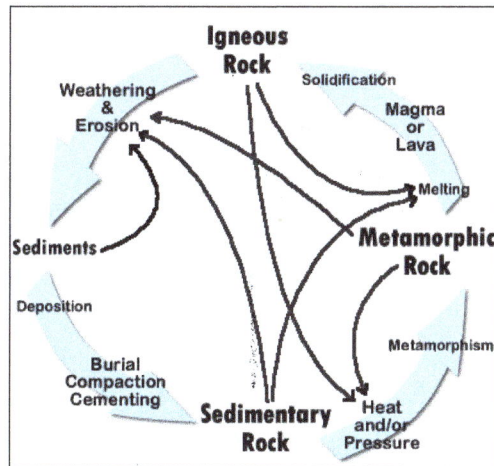

The rock cycle.

- Sedimentary rocks are formed via weathering and transport of existing rocks, and then deposition, cementation, and compaction into a sedimentary rock.

- Igneous rocks are formed by cooling and crystallization of molten rock.

- Metamorphic rocks are formed when heat or pressure are applied to other rocks.

The primary agent driving these processes is the movement of Earth's tectonic plates, which creates mountains, volcanoes, and ocean basins. Changes in the rate that rocks are made and destroyed can have a profound effect on the planet. As the rate of plate tectonic movements has changed over geologic time scales, the rock cycle has changed as well, and these changes have been able to affect climate. For example, at times when the rate of plate movements has been high, there is more volcanic activity, which releases more particles into the atmosphere. Faster plate tectonic movements also mean more mountains are built in areas where plates converge. As rocks are uplifted into mountains, they start to erode and dissolve, sending sediments and nutrients into waterways and impacting the ecosystems for living things.

Soil Chemistry

Soil chemistry is discipline which embracing all chemical and mineralogical compounds and reactions occurring in soils and soil-forming processes. The goals of soil chemistry are: (1) to establish, through chemical analysis, compositional limits of natural soil types and optimal growth conditions for the various plant communities, (2) to derive principles governing development of soil profiles, and (3) to apply analytical and theoretical knowledge to the improvement of agricultural production.

Soils transport and move water, provide homes for thousands of bacteria and other creatures, and have many different arrangements of weathered rock and minerals. When soils and minerals weather over time, the chemical composition of soil also changes. However, nothing changes the chemistry of soils faster than humans do.

Many of today's soil chemistry problems have to do with environmental sciences.

Ion Exchanges

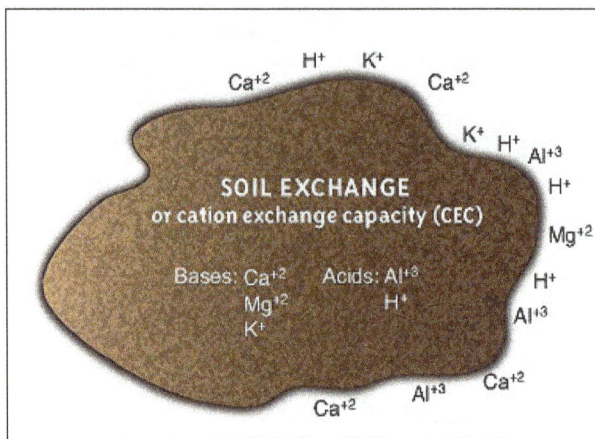

This diagram represents soil cations attached firmly to the soil.

Ion exchange involves the movement of cations (positively charged elements like calcium, -magnesium, and sodium) and anions (negatively charged elements like chloride, and compounds like nitrate) through the soils. In the United States, cation exchange is much more common.

Cation exchange is the interchanging between a cation in the solution of water around the soil particle, and another cation that is stuck to the clay surface. The number of cations in the soil water solution is much smaller than the number that is attached to soil particles.

The total amount of positive charges that the soil can absorb is called the cation exchange capacity (CEC). CEC impacts how quickly nutrients move through the profile. A soil with a low CEC is much less fertile because it cannot hold on to many nutrients, and they usually contain less clays. If your soil has a low CEC, it is important to apply fertilizer small doses so it does not infiltrate into the groundwater. A soil with a low CEC is less able to hold spilt chemicals.

Soil pH

The soil pH is a measure of soil acidity or alkalinity. pH can range from 1 to 14, with values 0-7 being acidic, and 7-14 being alkaline. Soils usually range from 4 to 10. The pH is one of the most important properties involved in plant growth, as well as understanding how rapidly reactions occur in the soil. .For example, the element iron becomes less available to plants a higher the pH is. This creates iron deficiency problems. Crops usually prefer values between 5.5-8, but the value depends on the crop. The pH of soil comes from the parent material during soil formation, but humans can add things to soils to change them to better suit plant growth. Soil pH also affects organisms.

Sorption and Precipitation

Soil particles have the ability to capture different nutrients and ions. Sorption is the process in which one substance takes up or holds another. In this case, soils that have high sorption can hold a lot of extra environmental contaminents, like phosporus, onto the particles. Soilprecipitation occurs during chemical reactions when a nutrient or chemical in the soil solution (water around soil particles) transforms into a solid. This is really important if soils are really salty. Soil chemists study the speed of these reactions under many different conditions.

Soil Organic Matter Interactions

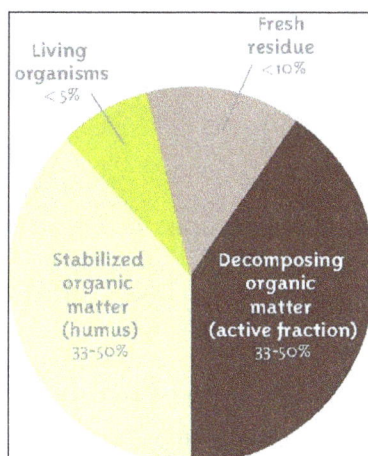

Soil has four major categories of organic matter inside of it, including active and long term types.

Soil chemists also study soil organic matter(OM), which are materials derived from the decay of plants and animals. They contain many hydrogen and carbon compounds. The arrangement and formation of these compounds influence a soils ability to handle spilt chemicals and other pollutants.

Oxidation and Reduction Reactions

Soils that alternate between wet and dry go from having a lot of oxygen to not a lot of oxygen. The presence or absence of oxygen determines how soils chemically react.Oxidation is the loss of electrons, and reduction is the gaining of electrons at the soil surface. These type of reactions occur every day, and are responsible for creating things like rust. Soils, because they contain a lot of iron, can also rust, or if they contain a lot of water, can turn a light gray color. This is partially responsible for all of the different colors that are found, and creates the speckles usually found deeper in the soil.

References

- Atmosphere: businessdictionary.com, Retrieved 05 June, 2019
- Atmospheric-chemistry: newworldencyclopedia.org, Retrieved 25 April, 2019
- What-is-the-atmosphere- 54760: universetoday.com, Retrieved 13 May, 2019
- Air-composition-properties: byjus.com, Retrieved 23 July, 2019
- Components-of-air, air-around-us: toppr.com, Retrieved 04 March, 2019
- Hydrosphere: energyeducation.ca, Retrieved 13 February, 2019
- Water-Chemistry, Environmental-Chemistry-Supplemental-Modules-(Environmental-Chemistry): chem.libretexts.org, Retrieved 19 March, 2019
- Continental-lithosphere, earth-and-planetary-sciences: sciencedirect.com, Retrieved 14 January, 2019

Biogeochemical Cycles

Biogeochemical cycles refer to the pathways concerned with the cycling of chemical substances through the biotic and abiotic atmosphere of the Earth. Oxygen cycle, nitrogen cycle, sulfur cycle, phosphorus cycle, carbon cycle, etc. are some of the cycles that fall under its domain. This chapter discusses all these biogeochemical cycles in detail.

The Earth is a closed system for matter, except for small amounts of cosmic debris that enter the Earth's atmosphere. This means that all the elements needed for the structure and chemical processes of life come from the elements that were present in the Earth's crust when it was formed billions of years ago. This matter, the building blocks of life, continually cycle through Earth's systems, the atmosphere, hydrosphere, biosphere, and lithosphere, on time scales that range from a few days to millions of years. These cycles are called biogeochemical cycles, because they include a variety of biological, geological, and chemical processes.

Many elements cycle through ecosystems, organisms, air, water, and soil. Many of these are trace elements. Other elements, including carbon, nitrogen, oxygen, hydrogen, sulfur, and phosphorus are critical components of all biological life. Together, oxygen and carbon account for 80 percent of the weight of human beings. Because these elements are key components of life, they must be available for biological processes. Carbon, however, is relatively rare in the Earth's crust, and nitrogen, though abundant in the atmosphere, is in a form that is not useable by living organisms. The biogeochemical cycles transport and store these important elements so that they can be used by living organisms. Each cycle takes many different pathways and has various reservoirs, or storage places, where elements may reside for short or long periods of time. Each of the chemical, biological, and geological processes varies in their rates of cycling. Some molecules may cycle very quickly depending on the pathway. Carbon atoms in deep ocean sediments may take hundreds to millions of years to cycle completely through the system. An average water molecule resides in the atmosphere for about ten days, although it may be transported many miles before it falls back to the Earth as rain.

How fast substances cycle depends on its chemical reactivity and whether or not it can be found in a gaseous state. A gaseous phase allows molecules to be transported quickly. Phosphorous has no gaseous phase and is relatively unreactive, so it moves very slowly through its cycle. Phosphorus is stored in large amounts in sediment in the oceans or in the Earth's crust and is recycled back to the surface only over very long periods of time through upwelling of ocean waters or weathering of rocks.

Biogeochemical cycles are subject to disturbance by human activities. Humans accelerate natural biogeochemical cycles when elements are extracted from their reservoirs, or sources, and deposited back into the environment (sinks). For example, humans have significantly altered the carbon cycle by extracting and combusting billions of tons of hydrocarbons in fossil that were buried deep in the Earth's crust, in addition to clearing vegetation that stores carbon. Global release of carbon through human activities has increased from 1 billion tons per year in 1940 to 6.5 billion tons per year in 2000. About half of this extra carbon is taken up by plants and the oceans, while the other half remains in the atmosphere.

Oxygen Cycle

Oxygen cycle is the circulation of oxygen in various forms through nature. Free in the air and dissolved in water, oxygen is second only to nitrogen in abundance among uncombined elements in the atmosphere. Plants and animals use oxygen to respire and return it to the air and water as carbon dioxide (CO_2). CO_2 is then taken up by algae and terrestrial green plants and converted into carbohydrates during the process of photosynthesis, oxygen being a by-product. The waters of the world are the main oxygen generators of the biosphere; their algae are estimated to replace about 90 percent of all oxygen used. Oxygen is involved to some degree in all the other biogeochemical cycles. For example, over time, detritus from living organisms transfers oxygen-containing compounds such as calcium carbonates into the lithosphere.

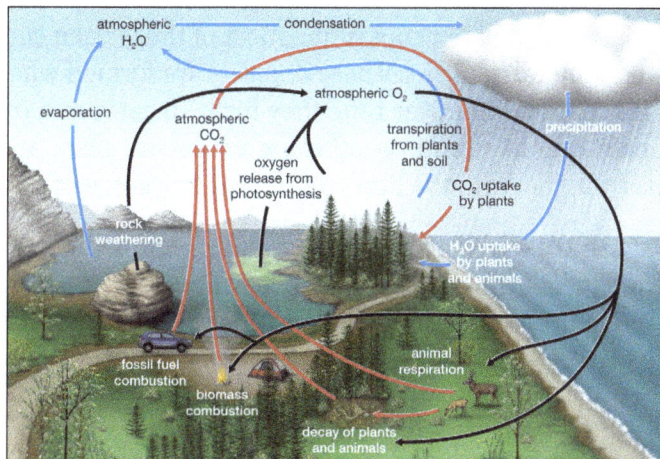

The generalized oxygen cycle.

Despite the burning of fossil fuel and the reduction of natural vegetation (on land and in the sea), the level of atmospheric oxygen appears to be relatively stable because of the increase in plant productivity resulting from agricultural advances worldwide.

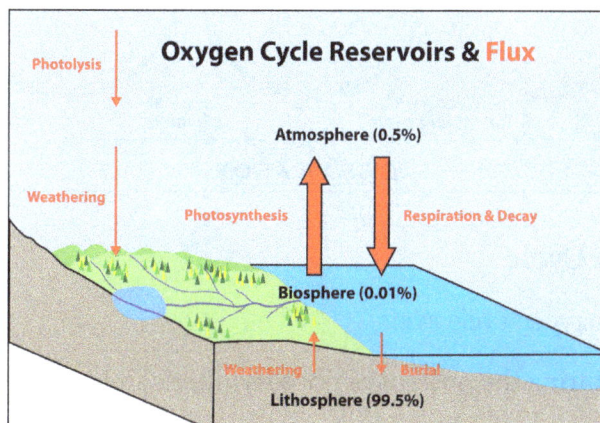

Oxygen Cycle Steps:

- Atmosphere: Only a small percentage of the world's oxygen is present in the atmosphere, only about 0.35 %. This exchange of gaseous oxygen happens through Photolysis.

  ◦ Photolysis: This is the process by which molecules like atmospheric water and nitrous oxide are broken down by the ultraviolet radiation coming from the sun and release free oxygen.

- Biosphere: The exchange of oxygen between the living beings on the planet, between the animal kingdom and the plant kingdom. The exchange of oxygen in the biosphere is codependent on the Carbon cycle and hydrogen cycle as well. It mainly occurs through 2 processes:

  ◦ Photosynthesis: The process by which plants make energy by taking in carbon dioxide from the atmosphere and give out oxygen.

  ◦ Respiration: The process by which animals and humans take in oxygen from the atmosphere and use it to break down carbohydrates and give out carbon dioxide.

- Lithosphere: The part of the planet containing most of the oxygen content through biomass, organic content and mineral deposits. These deposits are formed when free radical elements were exposed to free oxygen and over time they form silicates and oxides. This trapped oxygen is released back due to several weathering processes. Also, animals and plants draw nutrient materials from the from the lithosphere and free some of the trapped oxygen.

- Hydrosphere: Oxygen dissolved in water is responsible for the sustenance of the aquatic ecosystem present beneath the surface. The hydrosphere is 33% oxygen by volume present mainly as a component of water molecules with dissolved molecules including carbonic acids and free oxygen.

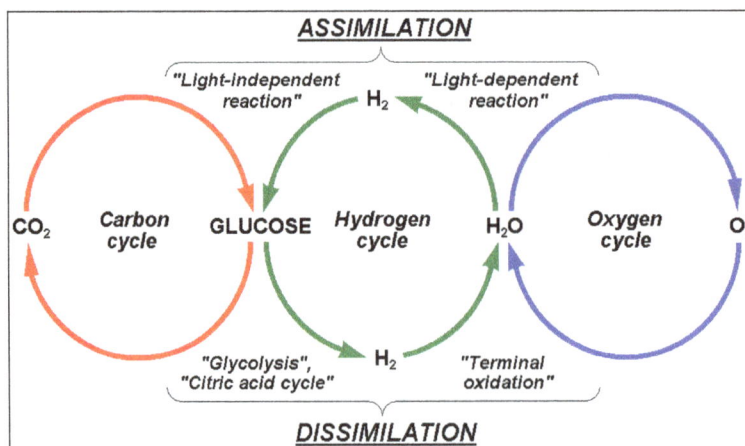

Stages of the Oxygen Cycle

The steps involved in the oxygen cycle are:

Stage-1: All green plants during the process of photosynthesis, release oxygen back into the atmosphere as a by-product.

Stage-2: All aerobic organisms use free oxygen for respiration.

Stage-3: Animals exhale Carbon dioxide back into the atmosphere which is again used by the plants during photosynthesis. Now oxygen is balanced within the atmosphere.

Nitrogen Cycle

Nitrogen is both the most abundant element in the atmosphere and, as a building block of proteins and nucleic acids such as DNA, a crucially important component of all biological life. The nitrogen cycle is a complex biogeochemical cycle in which nitrogen is converted from its inert atmospheric molecular form (N_2) into a form that is useful in biological processes.

Nitrogen cycle consists of four main steps namely:

- Nitrogen Fixation.
- Ammonification/Decay.
- Nitrification.
- De-nitrification.

It is important to note that microorganisms play an important role in each of these steps.

Mechanism of Each Step

Nitrogen Fixation

This is the first step of the nitrogen cycle. This step is characterized by the conversion of atmospheric N_2 into ammonia (NH_3). Bacteria like Azotobacter and Rhizobiumhave a major role in this process. They are harbored in the roots of the leguminous plants and help convert inert nitrogen to ammonia. Nitrogen fixation can occur in any of the following ways: atmospheric fixation (involves lightening), industrial fixation(manufacturing ammonia under high temperature and pressure condition).

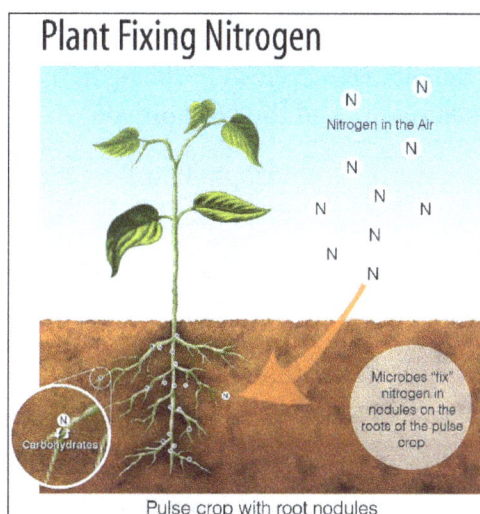

Plant Fixing Nitrogen

Nitrogen in the Air

Microbes "fix" nitrogen in nodules on the roots of the pulse crop.

Carbohydrates

Pulse crop with root nodules

Assimilation

Once the nitrogen has been fixed in the soil, plants can absorb nitrogen through their roots. This process of absorption is known as assimilation.

Ammonification

This is another process by which ammonia can be generated. Organic remains of plants and animals are broken down in the soil by some bacteria to release ammonia into the soil. This dead and waste matter is used by these microorganisms as food and they release ammonia into the soil.

Nitrification

This occurs in two-steps. The first step is in which NH3/NH$+ is converted to NO3- (nitrates). The bacteria Nitrosomonas and Nitrococcus present in the soil convert NH3 to NO2-, and another bacterium, Nitrobacter converts NO_2^- to NO_3^-. These bacteria gain energy through these conversions.

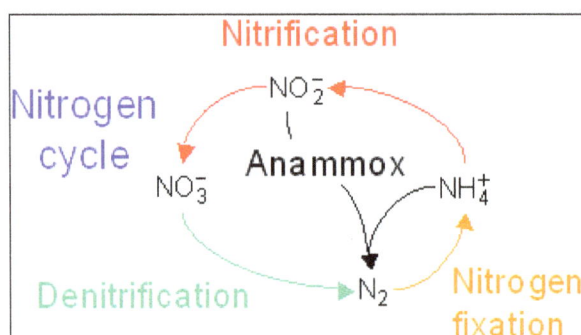

De-nitrification

Is the reverse of nitrification that occurs in the deep layers of soil where the bacteria convert NO_3^- is converted into N_2 and other gaseous compounds like NO_2. This occurs because in deep layers of soil, oxygen is not available and the soil bacteria use these nitrogen compounds instead of oxygen.

Importance of the Nitrogen Cycle

- As we all know by now, the nitrogen cycle helps bring in the inert nitrogen from the air into the biochemical process in plants and then to animals.

- Plants need nitrogen to synthesize chlorophyll and so the nitrogen cycle is absolutely essential for them.

- During the process of ammonification, the bacteria help degrade decomposing animal and plant matter. This helps in naturally cleaning up the environment.

- Due to the nitrogen cycle, nitrates and nitrites are released into the soil which helps in enriching the soil with nutrients needed for cultivation.

- As plants use nitrogen for their biochemical processes, animals obtain the nitrogen and nitrogen compounds from plants. Nitrogen is needed as is an integral part of the cell composition. It is due to the nitrogen cycle that animals are also able to utilize the nitrogen present in the air.

Bacteria play a Key Role in the Nitrogen Cycle

Nitrogen enters the living world by way of bacteria and other single-celled prokaryotes, which convert atmospheric nitrogen— N_2—into biologically usable forms in a process called nitrogen fixaon. Some species of nitrogen-fixing bacteria are free-living in soil or water, while others are beneficial symbionts that live inside of plants.

Nitrogen-fixing microorganisms capture atmospheric nitrogen by converng it to ammonia—N_3 — which can be taken up by plants and used to make organic molecules. The nitrogen-containing molecules are passed to animals when the plants are eaten. They may be incorporated into the animal's body or broken down and excreted as waste, such as the urea found in urine.

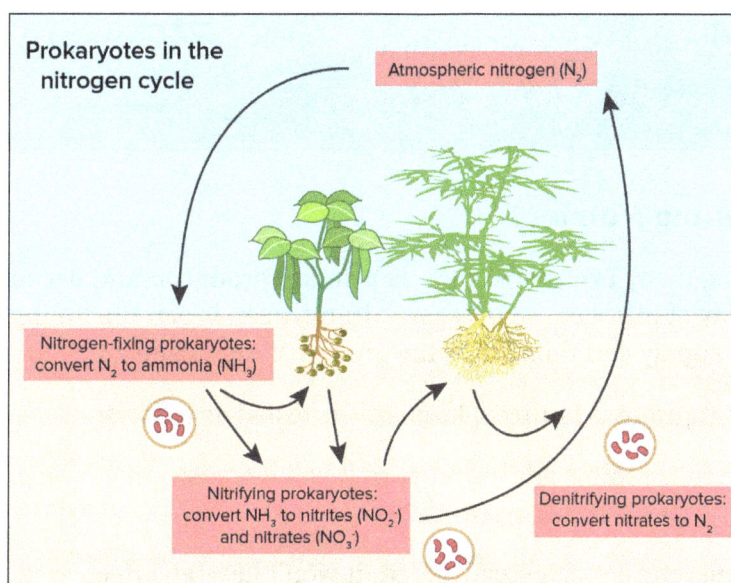

Nitrogen doesn't remain forever in the bodies of living organisms. Instead, it's converted from organic nitrogen back into N_2 gas by bacteria. This process often involves several steps in terrestrial—land—ecosystems. Nitrogenous compounds from dead organisms or wastes are converted into ammonia— NH_3—by bacteria, and the ammonia is converted into nitrites and nitrates. In the end, the nitrates are made into gas by denitrifying prokaryotes.

Nitrogen Cycling in Marine Ecosystems

So far, we've focused on the natural nitrogen cycle as it occurs in terrestrial ecosystems. However, generally similar steps occur in the marine nitrogen cycle. There, the ammonificaon, nitrificaon, and denitrification processes are performed by marine bacteria and archaea.

Some nitrogen-containing compounds fall to the ocean floor as sediment. Over long periods of me, the sediments get compressed and form sedimentary rock. Eventually, geological uplift may move the sedimentary rock to land. In the past, scienstists did not think that this nitrogen rich sedimentary rock was an important nitrogen source for terrestrial ecosystems. However, a new study suggests that it may actually be quite important—the nitrogen is released gradually to plants as the rock wears away, or weathers.

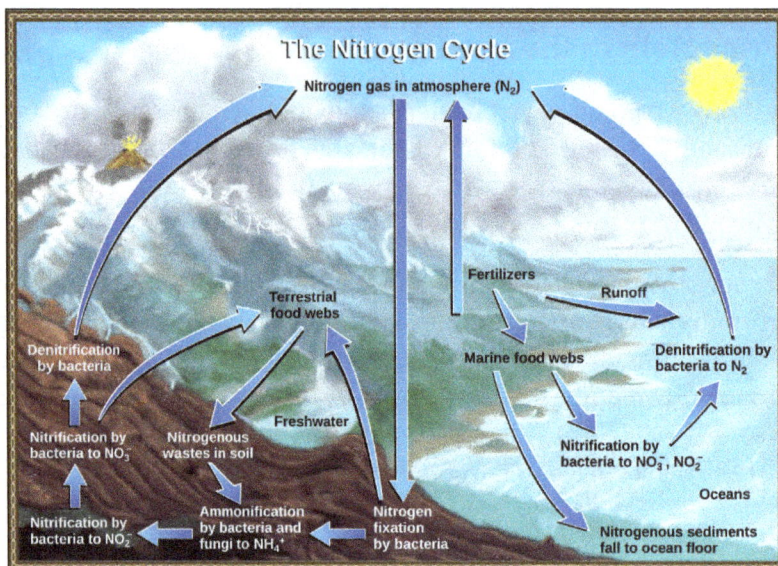

Nitrogen as a Liming Nutrient

In natural ecosystems, many processes, such as primary producon and decomposion, are limited by the available supply of nitrogen. In other words, nitrogen is oen the liming nutrient, the nutrient that's in shortest supply and thus limits the growth of organisms or populations.

How do we know if a nutrient is liming? Often, this is tested as follows:

*   When a nutrient is liming, adding more of it will increase growth—e.g., it will cause plants to grow taller than if nothing were added.

*   If a non-liming nutrient is instead added, it won't have an effect—e. g., plants will grow to the same height whether the nutrient is present or absent.

For example, if we added nitrogen to half the bean plants in a garden and found that they grew taller than untreated plants, that would suggest nitrogen was liming. If, instead, we didn't see a difference in growth in our experiment, that would suggest that some other nutrient than nitrogen must be limiting.

Nitrogen and phosphorus are the two most common liming nutrients in both natural ecosystems and agriculture. That's why, if you look at a bag of fertilizer, you will see it contains a lot of nitrogen and phosphorus.

Human Activity Affects Cycling of Nitrogen

We humans may not be able to fix nitrogen biologically, but we certainly do industrially! About 450 million metric tons of fixed nitrogen are made each year using a chemical method called the Haber-Bosch process, in which N_2 is reacted with hydrogen—H_2—at high temperatures. Most of this fixed nitrogen goes to make ferlizers we use on our lawns, gardens, and agricultural fields.

In general, human acvity releases nitrogen into the environment by two main means: combuson of fossil fuels and use of nitrogencontaining ferlizers in agriculture. Both processes increase levels of nitrogen-containing compounds in the atmosphere. High levels of atmospheric nitrogen—other than N_2—are associated with harmful effects, like the producon of acid rain—as nitric acid, HNO_3—and contribuons to the greenhouse effect—as nitrous oxide, H_2O.

Also, when artificial fertilizers containing nitrogen and phosphorus are used in agriculture, the excess fertilizer may be washed into lakes, streams, and rivers by surface runoff. A major effect from fertilizer runoff is saltwater and freshwater eutrophication. In this process, nutrient runoff causes overgrowth, or a "bloom," of algae or other microorganisms. Without the nutrient runoff, they were limited in their growth by availability of nitrogen or phosphorus.

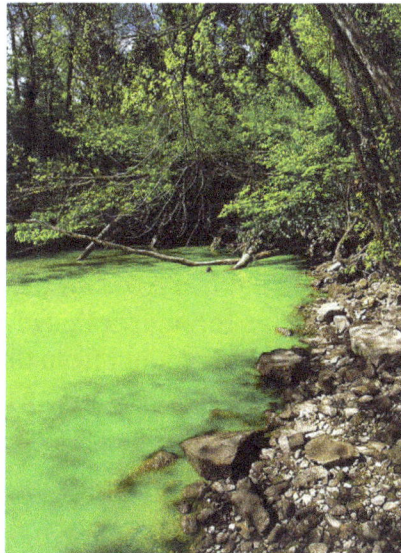

Eutrophication can reduce oxygen availability in the water during the night time because the algae and microorganisms that feed on them use up large quantities of oxygen in cellular respiration. This can cause the death of other organisms living in the affected ecosystems, such as fish and shrimp, and result in low-oxygen, species-depleted areas called dead zones.

Sulfur Cycle

Sulfur (S) is the most abundant element in the universe, is a brittle, yellow, tasteless, and odorless non-metallic element. It comprises many vitamins, proteins, and hormones that play critical roles in both climate and in the health of various ecosystems. The majority of the Earth's sulfur is stored underground in rocks and minerals, including as sulfate salts buried deep within ocean sediments.

The sulfur cycle contains both atmospheric and terrestrial processes. Within the terrestrial portion, the cycle begins with the weathering of rocks, releasing the stored sulfur. The sulfur then comes into contact with air where it is converted into sulfate (SO_4). The sulfate is taken up by plants and microorganisms and is converted into organic forms; animals then consume these organic forms through foods they eat, thereby moving the sulfur through the food chain. As organisms die and decompose, some of the sulfur is again released as a sulfate and some enters the tissues of micro-organisms. There are also a variety of natural sources that emit sulfur directly into the atmosphere, including volcanic eruptions, the breakdown of organic matter in swamps and tidal flats, and the evaporation of water.

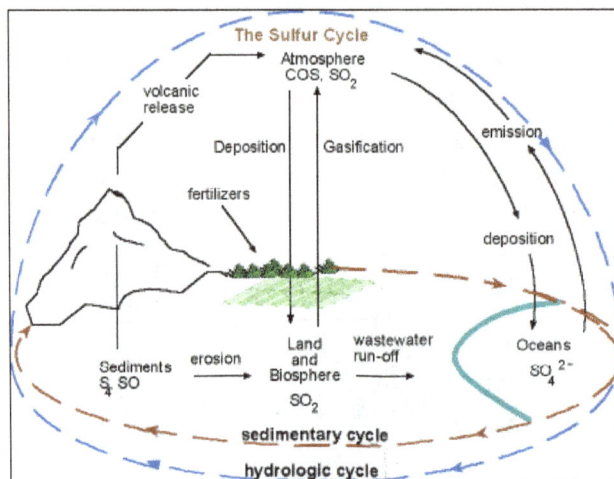

Sulfur eventually settles back into the Earth or comes down within rainfall. A continuous loss of sulfur from terrestrial ecosystem runoff occurs through drainage into lakes and streams, and eventually oceans. Sulfur also enters the ocean through fallout from the Earth's atmosphere. Within the ocean, some sulfur cycles through marine communities, moving through the food chain. A portion of this sulfur is emitted back into the atmosphere from sea spray. The remaining sulfur is lost to the ocean depths, combining with iron to form ferrous sulfide which is responsible for the black color of most marine sediments.

Since the Industrial Revolution, human activities have contributed to the amount of sulfur that enters the atmosphere, primarily through the burning of fossil fuels and the processing of metals. One-third of all sulfur that reaches the atmosphere—including 90% of sulfur dioxide—stems from human activities. Emissions from these activities, along with nitrogen emissions, react with other chemicals in the atmosphere to produce tiny particles of sulfate salts which fall as acid rain, causing a variety of damage to both the natural environment as well as to man-made environments, such as the chemical weathering of buildings. However, as particles and tiny airborne droplets, sulfur also acts as a regulator of global climate. Sulfur dioxide and sulfate aerosols absorb ultraviolet radiation, creating cloud cover that cools cities and may offset global warming caused by the greenhouse effect.

Sulfur Oxidation State

Sulfur has four main oxidation states in nature, which are -2, +2, +4, and +6. The common sulfur species of each oxidation state are listed as follows:

S^{2-}: H_2S, FeS, FeS_2, CuS

S_0: native, or elemental, sulfur

S^{2+}: SO

S^{4+}: SO_2, sulfite (SO_3^{2-})

S^{6+}: SO_4^{2-} (H_2SO_4, $CaSO_4$), SF_6

Sulfur Sources and Sinks

Sulfur is found in oxidation states ranging from +6 in SO_4^{2-} to -2 in sulfides. Thus, elemental sulfur can either give or receive electrons depending on its environment. On the anoxic early Earth, most sulfur was present in minerals such as pyrite (FeS_2). Over Earth history, the amount of mobile sulfur increased through volcanic activity as well as weathering of the crust in an oxygenated atmosphere. Earth's main sulfur sink is the oceans SO_4^{2-}, where it is the major oxidizing agent.

When SO_4^{2-} is assimilated by organisms, it is reduced and converted to organic sulfur, which is an essential component of proteins. However, the biosphere does not act as a major sink for sulfur, instead the majority of sulfur is found in seawater or sedimentary rocks including: pyrite rich shales, evaporite rocks (anhydrite and baryte), and calcium and magnesium carbonates (i.e. carbonate-associated sulfate). The amount of sulfate in the oceans is controlled by three major processes:

- Input from rivers;

- Sulfate reduction and sulfide re-oxidation on continental shelves and slopes;

- Burial of anhydrite and pyrite in the oceanic crust.

The primary natural source of sulfur to the atmosphere is sea spray or windblown sulfur rich dust, neither of which is long lived in the atmosphere. In recent times, the large annual input of sulfur from the burning of coal and other fossil fuels has added a substantial amount SO_2 which acts as an air pollutant. In the geologic past, igneous intrusions into coal measures have caused large scale burning of these measures, and consequential release of sulfur to the atmosphere. This has led to substantial disruption to the climate system, and is one of the proposed causes of the Permian–Triassic extinction event.

Dimethylsulfide [$(CH_3)_2S$ or DMS] is produced by the decomposition of dimethylsulfoniopropionate (DMSP) from dying phytoplankton cells in the ocean's photic zone, and is the major biogenic gas emitted from the sea, where it is responsible for the distinctive "smell of the sea" along coastlines. DMS is the largest natural source of sulfur gas, but still only has a residence time of about one day in the atmosphere and a majority of it is redeposited in the oceans rather than making it to land. However, it is a significant factor in the climate system, as it is involved in the formation of clouds.

Biologically and Thermochemically Driven Sulfate Reduction

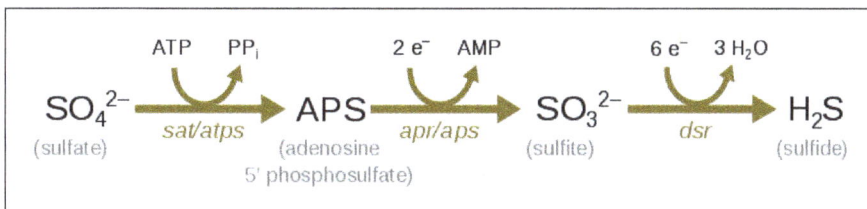

Dissimilatory sulfate reduction pathway.

Through the dissimilatory sulfate reduction pathway, sulfate can be reduced either bacterially (bacterial sulfate reduction) or inorganically (thermochemical sulfate reduction). This pathway involves the reduction of sulfate by organic compounds to produce hydrogen sulfide, which occurs in both processes.

The main products and reactants of bacterial sulfate reduction (BSR) and thermochemical sulfate reduction (TSR) are very similar. For both, various organic compounds and dissolved sulfate are the reactants, and the products or by-products are as follows: H_2S, CO_2, carbonates, elemental sulfur and metal sulfides. However, the reactive organic compounds differ for BSR and TSR because of the mutually exclusive temperature regimes. Organic acids are the main organic reactants for BSR and branched/n-alkanes are the main organic reactants for TSR. The inorganic reaction products in BSR and TSR are $H_2S(HS^-)$ and HCO_3^- (CO_2).

These processes occur because there are two very different thermal regimes in which sulfate is reduced, particularly in low-temperature and high-temperature environments. BSR usually occurs at lower temperatures from 0-80°C, while TSR happens at much higher temperatures around 100–140°C. Temperatures for TSR are not as well defined; the lowest confirmed temperature is 127°C and the highest temperatures occur in settings around 160-180°C. These two different regimes appear because at higher temperatures most sulfate-reducing microbes can no longer metabolize due to the denaturation of proteins or deactivation of enzymes, so TSR takes over. However, in hot sediments around hydrothermal vents BSR can happen at temperatures up to 110°C.

BSR and TSR occur at different depths. BSR takes place in low-temperature environments, which are shallower settings such as oil and gas fields. BSR can also take place in modern marine sedimentary environments such as stratified inland seas, continental shelves, organic-rich deltas, and hydrothermal sediments which have intense microbial sulfate reduction because of the high concentration of dissolved sulfate in the seawater. Additionally, the high amounts of hydrogen sulfide found in oil in gas fields is thought to arise from the oxidation of petroleum hydrocarbons by sulfate. Such reactions are known to occur by microbial processes but it is generally accepted that TSR is responsible for the bulk of these reactions, especially in deep or hot reservoirs. Thus, TSR occurs in deep reservoirs where the temperatures are much higher. BSR is geologically instantaneous in most geologic settings, while TSR occurs at rates in the order of hundreds of thousands of years. Although much slower than BSR, even TSR appears to be a geologically fairly fast process.

BSR in shallow environments and TSR in deep reservoirs are key processes in the oceanic sulfur cycle. Approximately, 10% (of the total gas) of H_2S is produced in BSR settings, whereas 90% of the H_2S is produced in TSR settings. If there is more than a few percent of H_2S in any deep reservoir, then it is assumed that TSR has taken over. This is due to the fact that thermal cracking of hydrocarbons doesn't provide more than a couple percent of H_2S. The amount of H_2S is affected by several factors such as, the availability of organic reactants and sulfate and the presence/availability of base and transition metals.

Sulfur-oxidizing Bacteria in Hydrothermal Vents

Hydrothermal vents emit hydrogen sulfide that support the carbon fixation of chemolithotrophic bacteria that oxidize hydrogen sulfide with oxygen to produce elemental sulfur or sulfate. The chemical reactions are as follows:

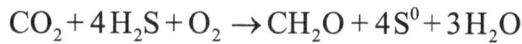

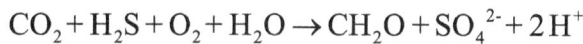

$$CO_2 + 4H_2S + O_2 \rightarrow CH_2O + 4S^0 + 3H_2O$$

$$CO_2 + H_2S + O_2 + H_2O \rightarrow CH_2O + SO_4^{2-} + 2H^+$$

In modern oceans, *Thiomicrospira*, *Halothiobacillus*, and *Beggiatoa* are primary sulfur oxidizing bacteria, and form chemosynthetic symbioses with animal hosts. The host provides metabolic substrates (e.g., CO_2, O_2, H_2O) to the symbiont while the symbiont generates organic carbon for sustaining the metabolic activities of the host. The produced sulfate usually combines with the leached calcium ions to form gypsum, which can form widespread deposits on near mid-ocean spreading centers.

$\delta^{34}S$

Although 25 isotopes are known for sulfur, only four are stable and of geochemical importance. Of those four, two (^{32}S, light and ^{34}S, heavy) comprise (99.22%) of S on Earth. The vast majority (95.02%) of S occurs as ^{32}S with only 4.21% in ^{34}S. The ratio of these two isotopes is fixed in our solar system and has been since its formation. The bulk Earth sulfur isotopic ratio is thought to be the same as the ratio of 22.22 measured from the Canyon Diablo troilite (CDT), a meteorite. That ratio is accepted as the international standard and is therefore set at δ0.00. Deviation from 0.00 is expressed as the $\delta^{34}S$ which is a ratio in per mill (‰). Positive values correlate to increased levels of ^{34}S, whereas negative values correlate with greater ^{32}S in a sample.

Formation of sulfur minerals through non-biogenic processes does not substantially differentiate between the light and heavy isotopes, therefore sulfur isotope ratios in gypsum or barite should be the same as the overall isotope ratio in the water column at their time of precipitation. Sulfate reduction through biologic activity strongly differentiates between the two isotopes because of the more rapid enzymic reaction with ^{32}S. Sulfate metabolism results in an isotopic depletion of -18‰, and repeated cycles of oxidation and reduction can result in values up to -50 ‰. Average present day seawater values of $\delta^{34}S$ are on the order of +21‰.

Throughout geologic history the sulfur cycle and the isotopic ratios have coevolved with the biosphere becoming overall more negative with the increases in biologically driven sulfate reduction, but also show substantial positive excursion. In general positive excursions in the sulfur isotopes mean that there is an excess of pyrite deposition rather than oxidation of sulfide minerals exposed on land.

Marine Sulfur Cycle

The sulfur cycle in marine environments has been well-studied via the tool of sulfur isotope systematics expressed as $\delta^{34}S$. The modern global oceans have sulfur storage of 1.3×10^{21} g, mainly occurring as sulfate with the $\delta^{34}S$ value of +21‰. The overall input flux is 1.0×10^{14} g/year with the sulfur isotope composition of ~3‰. Riverine sulfate derived from the terrestrial weathering of sulfide minerals ($\delta^{34}S = +6‰$) is the primary input of sulfur to the oceans. Other sources are metamorphic and volcanic degassing and hydrothermal activity ($\delta^{34}S = 0‰$), which release reduced sulfur species (e.g., H_2S and S^0). There are two major outputs of sulfur from the oceans. The first sink is the burial of sulfate either as marine evaporites (e.g., gypsum) or carbonate-associated sulfate (CAS), which accounts for 6×10^{13} g/year ($\delta^{34}S$ = +21‰). The second sulfur sink is pyrite burial in shelf sediments or deep seafloor sediments (4×10^{13} g/year; $\delta^{34}S$ = -20‰). The total marine sulfur output flux is 1.0×10^{14} g/year which matches the input fluxes, implying the modern marine sulfur budget is at steady state. The residence time of sulfur in modern global oceans is 13,000,000 years.

Evolution of the Sulfur Cycle

The isotopic composition of sedimentary sulfides provides primary information on the evolution of the sulfur cycle.

The total inventory of sulfur compounds on the surface of the Earth (nearly 10^{22} g S) represents the total outgassing of sulfur through geologic time. Rocks analyzed for sulfur content are generally organic-rich shales meaning they are likely controlled by biogenic sulfur reduction. Average seawater curves are generated from evaporites deposited throughout geologic time because again, since they do not discriminate between the heavy and light sulfur isotopes, they should mimic the ocean composition at the time of deposition.

4.6 billion years ago (Ga) the Earth formed and had a theoretical $\delta^{34}S$ value of 0. Since there was no biologic activity on early Earth there would be no isotopic fractionation. All sulfur in the atmosphere would be released during volcanic eruptions. When the oceans condensed on Earth, the atmosphere was essentially swept clean of sulfur gases, owing to their high solubility in water. Throughout the majority of the Archean (4.6–2.5 Ga) most systems appeared to be sulfate-limited. Some small Archean evaporite deposits require that at least locally elevated concentrations (possibly due to local volcanic activity) of sulfate existed in order for them to be supersaturated and precipitate out of solution.

3.8–3.6 Ga marks the beginning of the exposed geologic record because this is the age of the oldest rocks on Earth. Metasedimentary rocks from this time still have an isotopic value of 0 because the biosphere was not developed enough (possibly at all) to fractionate sulfur.

3.5 Ga anoxyogenic photosynthesis is established and provides a weak source of sulfate to the global ocean with sulfate concentrations incredibly low the $\delta^{34}S$ is still basically 0. Shortly after, at 3.4 Ga the first evidence for minimal fractionation in evaporitic sulfate in association with magmatically derived sulfides can be seen in the rock record. This fractionation shows possible evidence for anoxygenic phototrophic bacteria.

2.8 Ga marks the first evidence for oxygen production through photosynthesis. This is important because there cannot be sulfur oxidation without oxygen in the atmosphere. This exemplifies the coevolution of the oxygen and sulfur cycles as well as the biosphere.

2.7–2.5 Ga is the age of the oldest sedimentary rocks to have a depleted $\delta^{34}S$ which provide the first compelling evidence for sulfate reduction.

2.3 Ga sulfate increases to more than 1 mM; this increase in sulfate is coincident with the "Great Oxygenation Event", when redox conditions on Earth's surface are thought by most workers to have shifted fundamentally from reducing to oxidizing. This shift would have led to an incredible increase in sulfate weathering which would have led to an increase in sulfate in the oceans. The large isotopic fractionations that would likely be associated with bacteria reduction are produced for the first time. Although there was a distinct rise in seawater sulfate at this time it was likely still only less than 5–15% of present-day levels.

At 1.8 Ga, Banded iron formations (BIF) are common sedimentary rocks throughout the Archean and Paleoproterozoic; their disappearance marks a distinct shift in the chemistry of ocean water. BIFs have alternating layers of iron oxides and chert. BIFs only form if the water is allowed to supersaturate in dissolved iron (Fe^{2+}) meaning there cannot be free oxygen or sulfur in the water column because it would form Fe^{3+} (rust) or pyrite and precipitate out of solution. Following this supersaturation, the water must become oxygenated in order for the ferric rich bands to precipitate it must still be sulfur poor otherwise pyrite would form instead of Fe^{3+}. It has been hypothesized that BIFs formed during the initial evolution of photosynthetic organisms that had phases of population growth, causing over production of oxygen. Due to this over production they would poison themselves causing a mass die off, which would cut off the source of oxygen and produce a large amount of CO_2 through the decomposition of their bodies, allowing for another bacterial bloom. After 1.8 Ga sulfate concentrations were sufficient to increase rates of sulfate reduction to greater than the delivery flux of iron to the oceans.

Along with the disappearance of BIF, the end of the Paleoproterozoic also marks the first large scale sedimentary exhalative deposits showing a link between mineralization and a likely increase in the amount of sulfate in sea water. In the Paleoproterozoic the sulfate in seawater had increased to an amount greater than in the Archean, but was still lower than present day values. The sulfate levels in the Proterozoic also act as proxies for atmospheric oxygen because sulfate is produced mostly through weathering of the continents in the presence of oxygen. The low levels in the Proterozoic simply imply that levels of atmospheric oxygen fell between the abundances of the Phanerozoic and the deficiencies of the Archean.

750 million years ago (Ma) there is a renewed deposition of BIF which marks a significant change in ocean chemistry. This was likely due to snowball earth episodes where the entire globe including the oceans was covered in a layer of ice cutting off oxygenation. In the late Neoproterozoic high carbon burial rates increased the atmospheric oxygen level to >10% of its present-day value. In the Latest Neoproterozoic another major oxidizing event occurred on Earth's surface that resulted in an oxic deep ocean and possibly allowed for the appearance of multicellular life.

During the last 600 million years, seawater SO_4 has varied between +10 and +30‰ in $\delta^{34}S$, with an average value close to that of today. This coincides with atmospheric O_2 levels reaching something close to modern values around the Precambrian–Cambrian boundary.

Over a shorter time scale (ten million years) changes in the sulfur cycle are easier to observe and can be even better constrained with oxygen isotopes. Oxygen is continually incorporated into the sulfur cycle through sulfate oxidation and then released when that sulfate is reduced once again. Since different sulfate sources within the ocean have distinct oxygen isotopic values it may be possible to use oxygen to trace the sulfur cycle. Biological sulfate reduction preferentially selects lighter oxygen isotopes for the same reason that lighter sulfur isotopes are preferred. By studying oxygen isotopes in ocean sediments over the last 10 million years were able to better constrain the sulfur concentrations in sea water through that same time. They found that the sea level changes due to Pliocene and Pleistocene glacial cycles changed the area of continental shelves which then disrupted the sulfur processing, lowering the concentration of sulfate in the sea water. This was a drastic change as compared to preglacial times before 2 million years ago.

The Great Oxidation Event and Sulfur Isotope Mass-independent Fractionation

The Great Oxygenation Event (GOE) is characterized by the disappearance of sulfur isotope mass-independent fractionation (MIF) in the sedimentary records at around 2.45 billion years ago (Ga). The MIF of sulfur isotope ($\Delta^{33}S$) is defined by the deviation of measured $\delta^{33}S$ value from the $\delta^{33}S$ value inferred from the measured $\delta^{34}S$ value according to the mass dependent fractionation law. The Great Oxidation Event represented a massive transition of global sulfur cycles. Before the Great Oxidation Event, the sulfur cycle was heavily influenced by the ultraviolet (UV) radiation and the associated photochemical reactions, which induced the sulfur isotope mass-independent fractionation ($\Delta^{33}S \neq 0$). The preservation of sulfur isotope mass-independent fractionation signals requires the atmospheric O_2 lower than 10^{-5} of present atmospheric level (PAL). The disappearance of sulfur isotope mass-independent fractionation at ~2.45 Ga indicates that atmospheric pO_2 exceeded 10^{-5} present atmospheric level after the Great Oxygenation Event. Oxygen played an essential role in the global sulfur cycles after the Great Oxygenation Event, such as oxidative weathering of sulfides. The burial of pyrite in sediments in turn contributes to the accumulation of free O_2 in Earth's surface environment.

Economic Importance

Sulfur is intimately involved in production of fossil fuels and a majority of metal deposits because of its ability to act as an oxidizing or reducing agent. The vast majority of the major mineral deposits on Earth contain a substantial amount of sulfur including, but not limited to: sedimentary exhalative deposits (SEDEX), Carbonate-hosted lead-zinc ore deposits (Mississippi Valley-Type MVT) and porphyry copper deposits. Iron sulfides, galena and sphalerite will form as by-products of hydrogen sulfide generation, as long as the respective transition or base metals are present or transported to a sulfate reduction site. If the system runs out of reactive hydrocarbons economically viable elemental sulfur deposits may form. Sulfur also acts as a reducing agent in many natural gas reservoirs and generally ore forming fluids have a close relationship with ancient hydrocarbon seeps or vents.

Important sources of sulfur in ore deposits are generally deep-seated, but they can also come from local country rocks, sea water, or marine evaporites. The presence or absence of sulfur is one of the limiting factors on both the concentration of precious metals and its precipitation from solution. pH, temperature and especially redox states determine whether sulfides will precipitate. Most sulfide brines will remain in concentration until they reach reducing conditions, a higher pH or lower temperatures.

Ore fluids are generally linked to metal rich waters that have been heated within a sedimentary basin under the elevated thermal conditions typically in extensional tectonic settings. The redox conditions of the basin lithologies exert an important control on the redox state of the metal-transporting fluids and deposits can form from both oxidizing and reducing fluids. Metal-rich ore fluids tend to be by necessity comparatively sulfide deficient, so a substantial portion of the sulfide must be supplied from another source at the site of mineralization. Bacterial reduction of seawater sulfate or a euxinic (anoxic and H_2S-containing) water column is a necessary source of that sulfide. When present, the $\delta^{34}S$ values of barite are generally consistent with a seawater sulfate source, suggesting baryte formation by reaction between hydrothermal barium and sulfate in ambient seawater.

Once fossil fuels or precious metals are discovered and either burned or milled, the sulfur become a waste product which must be dealt with properly or it can become a pollutant. There has been a great increase in the amount of sulfur in our present day atmosphere because of the burning of fossil fuels. Sulfur acts as a pollutant and an economic resource at the same time.

Human Impact

Human activities have a major effect on the global sulfur cycle. The burning of coal, natural gas, and other fossil fuels has greatly increased the amount of S in the atmosphere and ocean and depleted the sedimentary rock sink. Without human impact sulfur would stay tied up in rocks for millions of years until it was uplifted through tectonic events and then released through erosion and weathering processes. Instead it is being drilled, pumped and burned at a steadily increasing rate. Over the most polluted areas there has been a 30-fold increase in sulfate deposition.

Although the sulfur curve shows shifts between net sulfur oxidation and net sulfur reduction in the geologic past, the magnitude of the current human impact is probably unprecedented in the geologic record. Human activities greatly increase the flux of sulfur to the atmosphere, some of which is transported globally. Humans are mining coal and extracting petroleum from the Earth's crust at a rate that mobilizes 150×10^{12} gS/yr, which is more than double the rate of 100 years ago. The result of human impact on these processes is to increase the pool of oxidized sulfur (SO_4) in the global cycle, at the expense of the storage of reduced sulfur in the Earth's crust. Therefore, human activities do not cause a major change in the global pools of S, but they do produce massive changes in the annual flux of S through the atmosphere.

When SO_2 is emitted as an air pollutant, it forms sulfuric acid through reactions with water in the atmosphere. Once the acid is completely dissociated in water the pH can drop to 4.3 or lower causing damage to both man-made and natural systems. According to the EPA, acid rain is a broad term referring to a mixture of wet and dry deposition (deposited material) from the atmosphere containing higher than normal amounts of nitric and sulfuric acids. Distilled water (water without any dissolved constituents), which contains no carbon dioxide, has a neutral pH of 7. Rain naturally has a slightly acidic pH of 5.6, because carbon dioxide and water in the air react together to form carbonic acid, a very weak acid. Around Washington, D.C., however, the average rain pH is between 4.2 and 4.4. Since pH is on a log scale dropping by 1 (the difference between normal rain water and acid rain) has a dramatic effect on the strength of the acid. In the United States, roughly 2/3 of all SO_2 and 1/4 of all NO_3 come from electric power generation that relies on burning fossil fuels, like coal.

Phosphorus Cycle

Phosphorus is an important element for all forms of life. As phosphate (PO_4), it makes up an important part of the structural framework that holds DNA and RNA together. Phosphates are also a critical component of ATP—the cellular energy carrier—as they serve as an energy release for organisms to use in building proteins or contacting muscles. Like calcium, phosphorus is important to vertebrates; in the human body, 80% of phosphorous is found in teeth and bones.

The phosphorus cycle differs from the other major biogeochemical cycles in that it does not include a gas phase; although small amounts of phosphoric acid (H_3PO_4) may make their way into the atmosphere, contributing—in some cases—to acid rain. The water, carbon, nitrogen and sulfur cycles all include at least one phase in which the element is in its gaseous state. Very little phosphorus circulates in the atmosphere because at Earth's normal temperatures and pressures, phosphorus and its various compounds are not gases. The largest reservoir of phosphorus is in sedimentary rock.

It is in these rocks where the phosphorus cycle begins. When it rains, phosphates are removed from the rocks (via weathering) and are distributed throughout both soils and water. Plants take up the phosphate ions from the soil. The phosphates then moves from plants to animals when herbivores eat plants and carnivores eat plants or herbivores. The phosphates absorbed by animal tissue through consumption eventually returns to the soil through the excretion of urine and feces, as well as from the final decomposition of plants and animals after death.

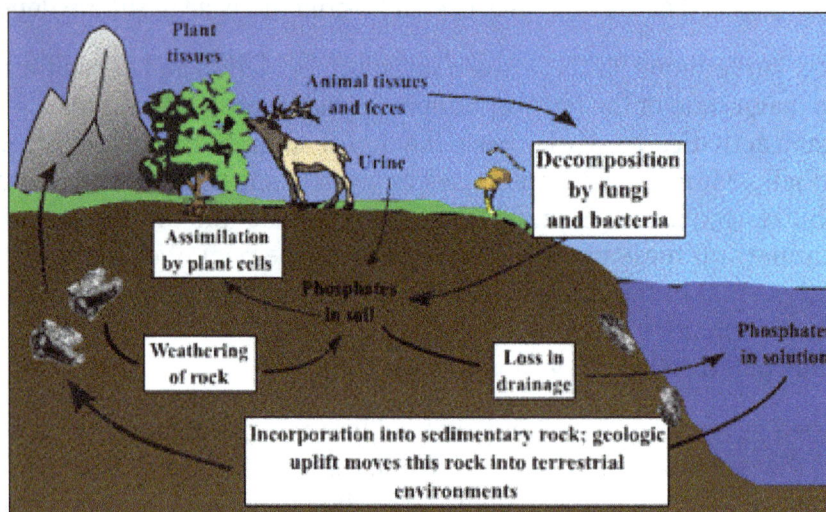

The same process occurs within the aquatic ecosystem. Phosphorus is not highly soluble, binding tightly to molecules in soil, therefore it mostly reaches waters by traveling with runoff soil particles. Phosphates also enter waterways through fertilizer runoff, sewage seepage, natural mineral deposits, and wastes from other industrial processes. These phosphates tend to settle on ocean floors and lake bottoms. As sediments are stirred up, phosphates may reenter the phosphorus cycle, but they are more commonly made available to aquatic organisms by being exposed through erosion. Water plants take up the waterborne phosphate which then travels up through successive stages of the aquatic food chain.

While obviously beneficial for many biological processes, in surface waters an excessive concentration of phosphorus is considered a pollutant. Phosphate stimulates the growth of plankton and plants, favoring weedy species over others. Excess growth of these plants tend to consume large amounts of dissolved oxygen, potentially suffocating fish and other marine animals, while also blocking available sunlight to bottom dwelling species. This is known as eutrophication.

Humans can alter the phosphorus cycle in many ways, including in the cutting of tropical rain forests and through the use of agricultural fertilizers. Rainforest ecosystems are supported primarily through the recycling of nutrients, with little or no nutrient reserves in their soils. As the forest is cut and/or burned, nutrients originally stored in plants and rocks are quickly washed away by heavy rains, causing the land to become unproductive. Agricultural runoff provides much of the phosphate found in waterways. Crops often cannot absorb all of the fertilizer in the soils, causing excess fertilizer runoff and increasing phosphate levels in rivers and other bodies of water. At one time the use of laundry detergents contributed to significant concentrations of phosphates in rivers, lakes, and streams, but most detergents no longer include phosphorus as an ingredient.

Steps of Phosphorus Cycle

Phosphorus Cycle goes through 4 main steps:

- Weathering.

- Absorption by Plants.

- Absorption by Animals.

- Return to the environment through decomposition.

Weathering

In light of the fact that phosphorous is most commonly found in rocks, the phosphorus cycle starts in the Earth's crust. Weathering causes phosphate salts to break from the rocks. The salts wash away into the ground, where they get mixed in the soil.

That takes us to the second step where phosphorous is taken up by plants and animals.

Absorption by Flora and Fauna

Plants absorb phosphate salts dissolved in soil water. It is worth noting that the quantities of phosphorus in the soil are normally small. This makes it one of the main factors that undermine plant growth.

That explains why we often apply phosphate fertilizers on agricultural land. Aquatic plants absorb inorganic form phosphorous from water and bottom layer of water bodies. In light of the fact that phosphate salts are not very water-soluble, they are also limiting factors for growth of plants in aquatic ecosystems.

Absorption by Animals

Animals absorb phosphorous by eating plants or plant-eating animals.

Namely, both marine and terrestrial plants are consumed by herbivorous animals, birds, and fish. Consequently, the organic form of phosphorus is transferred to the next level of consumers. Hence, the carnivores obtain phosphorous by eating herbivores.

It is worth noting that the rate of the phosphorus cycle is much faster in plants and animals than it is in rocks and sediments.

Return of Phosphorous back to the Ecosystem

When plants, animals, birds, and fish die, they are deposited in soil and water bodies. Under conducive environmental conditions, bacteria and other microbes decompose the dead organisms. During decomposition, the organic form of phosphorous is converted into an inorganic form, which is then recycled to the soil and water.

After that, the element will end up in rock formations or sediments, where it will remain for millions of years. In the end, phosphorous is released again to the soil through weathering and absorbed by the plants. And the phosphorus cycle starts over.

Phosphorous stays in a particular ecosystem unit for a longer period, but it eventually recycles back to the environment. Hence, the percentage composition of this element remains at a virtually constant value.

Eutrophication and Dead Zones

Most fertilizers used in agriculture—and on lawns and gardens—contain both nitrogen and phosphorus, which may be carried to aquatic ecosystems in surface runoff. Fertilizer carried in runoff may cause excessive growth of algae or other microbes that were previously limited by nitrogen or phosphorus. This phenomenon is called *eutrophication*. At least in some cases, phosphorus, not nitrogen, seems to be the main driver of eutrophication.

Some algae make water taste or smell bad or produce toxic compounds. Also, when all of those algae die and are decomposed by microbes, large amounts of oxygen are used up as their bodies are broken down. This spike in oxygen usage can sharply lower dissolved oxygen levels in the water and may lead to death by hypoxia—lack of oxygen—for other aquatic organisms, such as shellfish and finfish.

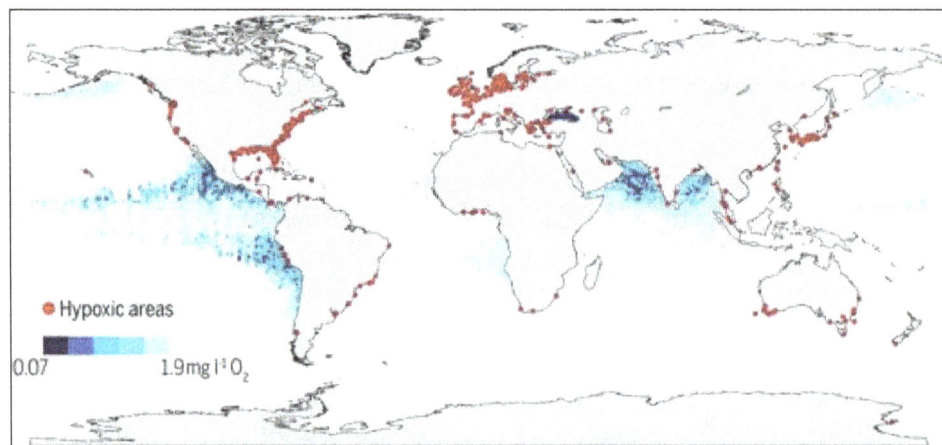

Regions of lakes and oceans that are depleted of oxygen due to a nutrient influx are called *dead zones*. The number of dead zones has increased for several years, and more than 400 of these zones existed in 2008. One of the worst dead zones is off the coast of the United States in the Gulf of Mexico. Fertilizer runoff from the Mississippi River Basin created a dead zone of over 8,463 square miles. As you can see in the figure below, dead zones are found in areas of high industrialization and population density around the world.

Fertilizers, phosphorus-containing detergents, and improperly disposed of sewage can all be sources of nitrogen and phosphorus that drive eutrophication. Using less fertilizer, eliminating phosphorus-containing detergents, and ensuring that sewage does not enter waterways—e.g., from a leaky septic system—are all ways that individuals, companies, and governments can help reduce eutrophication.

Carbon Cycle

There are a few types of atoms that can be a part of a plant one day, an animal the next day, and then travel downstream as a part of a river's water the following day. These atoms can be a part of both living things like plants and animals, as well as non-living things like water, air, and even rocks. The same atoms are recycled over and over in different parts of the Earth. This type of cycle of atoms between living and non-living things is known as a biogeochemical cycle.

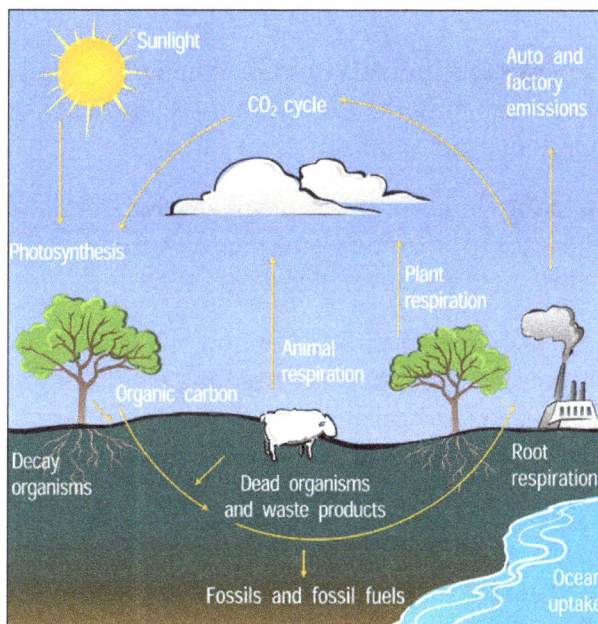

A simple diagram of parts of the carbon cycle, emphasizing the terrestrial (land-based) parts of the cycle.

All of the atoms that are building blocks of living things are a part of biogeochemical cycles. The most common of these are the carbon and nitrogen cycles.

Tiny atoms of carbon and nitrogen are able to move around the planet through these cycles. For example, an atom of carbon is absorbed from the air into the ocean water where it is used by little floating plankton doing photosynthesis to get the nutrition they need. There is the possibility that this little carbon atom becomes part of the plankton's skeleton, or a part of the skeleton of the larger animal that eats it, and then part of a sedimentary rock when the living things die and only bones are left behind. Carbon that is a part of rocks and fossil fuels like oil, coal, and natural gas may be held away from the rest of the carbon cycle for a long time. These long-term storage places are called "sinks". When fossil fuels are burned, carbon that had been underground is sent into the air as carbon dioxide, a greenhouse gas.

Recently, people have been causing these biogeochemical cycles to change. When we cut down forests, make more factories, and drive more cars that burn fossil fuels, the way that carbon and nitrogen move around the Earth changes. These changes add more greenhouse gases in our atmosphere and this causes climate change.

The element carbon is a part of seawater, the atmosphere, rocks such as limestone and coal, soils, as well as all living things. On our dynamic planet, carbon is able to move from one of these realms to another as a part of the carbon cycle.

- Carbon moves from the atmosphere to plants. In the atmosphere, carbon is attached to oxygen in a gas called carbon dioxide (CO_2). Through the process of photosynthesis, carbon dioxide is pulled from the air to produce food made from carbon for plant growth.

- Carbon moves from plants to animals. Through food chains, the carbon that is in plants moves to the animals that eat them. Animals that eat other animals get the carbon from their food too.

- Carbon moves from plants and animals to soils. When plants and animals die, their bodies, wood and leaves decays bringing the carbon into the ground. Some is buried and will become fossil fuels in millions and millions of years.

- Carbon moves from living things to the atmosphere. Each time you exhale, you are releasing carbon dioxide gas (CO_2) into the atmosphere. Animals and plants need to get rid of carbon dioxide gas through a process called respiration.

- Carbon moves from fossil fuels to the atmosphere when fuels are burned. When humans burn fossil fuels to power factories, power plants, cars and trucks, most of the carbon quickly enters the atmosphere as carbon dioxide gas. Each year, five and a half billion tons of carbon is released by burning fossil fuels. Of this massive amount, 3.3 billion tons stays in the atmosphere. Most of the remainder becomes dissolved in seawater.

- Carbon moves from the atmosphere to the oceans. The oceans, and other bodies of water, absorb some carbon from the atmosphere. The carbon is dissolved into the water.

Carbon dioxide is a greenhouse gas and traps heat in the atmosphere. Without it and other greenhouse gases, Earth would be a frozen world. But since the start of the Industrial Revolution about 150 years ago humans have burned so much fuel and released so much carbon dioxide into the air that global climate has risen over one degree Fahrenheit. The atmosphere has not held this much carbon for at least 420,000 years according to data from ice cores. The recent increase in amounts of greenhouse gases such as carbon dioxide is having a significant impact on the warming of our planet.

Carbon moves through our planet over longer time scales as well. For example, over millions of years weathering of rocks on land can add carbon to surface water which eventually runs off to the ocean. Over long time scales, carbon is removed from seawater when the shells and bones of marine animals and plankton collect on the sea floor. These shells and bones are made of limestone, which contains carbon. When they are deposited on the sea floor, carbon is stored from the rest of the carbon cycle for some amount of time. The amount of limestone deposited in the ocean depends somewhat on the amount of warm, tropical, shallow oceans on the planet because this is where prolific limestone-producing organisms such as corals live. The carbon can be released back to the atmosphere if the limestone melts or is metamorphosed in a subduction zone.

Biological and Geological Cycles

The Biological Carbon Cycle

Carbon enters all food webs, both terrestrial and aquatic, through autotrophs, or self-feeders. Almost all of these autotrophs are photosynthesizers, such as plants or algae.

Autotrophs capture carbon dioxide from the air or bicarbonate ions from the water and use them to make organic compounds such as glucose. Heterotrophs, or other-feeders, such as humans, consume the organic molecules, and the organic carbon is passed through food chains and webs.

To release the energy stored in carbon-containing molecules, such as sugars, autotrophs and heterotrophs break these molecules down in a process called cellular respiration. In this process, the carbons of the molecule are released as carbon dioxide. Decomposers also release organic compounds and carbon dioxide when they break down dead organisms and waste products.

Carbon can cycle quickly through this biological pathway, especially in aquatic ecosystems. Overall, an estimated 1,000 to 100,000 million metric tons of carbon move through the biological pathway each year. For context, a metric ton is about the weight of an elephant or a small car!

The Geological Carbon Cycle

The geological pathway of the carbon cycle takes much longer than the biological pathway described above. In fact, it usually takes millions of years for carbon to cycle through the geological pathway. Carbon may be stored for long periods of time in the atmosphere, bodies of liquid water— mostly oceans— ocean sediment, soil, rocks, fossil fuels, and Earth's interior.

The level of carbon dioxide in the atmosphere is influenced by the reservoir of carbon in the oceans and vice versa. Carbon dioxide from the atmosphere dissolves in water and reacts with water molecules in the following reactions:

$$CO_2 + H_2O \rightleftharpoons H_2CO_3 \rightleftharpoons HCO_3^- + H^+ \rightleftharpoons CO_3^{2-} + 2H^+$$

The carbonate $-CO_3^{2-}-$ released in this process combines with Ca^{2+} ions to make calcium carbonate

$-CaCO_3-$ a key component of the shells of marine organisms. When the organisms die, their remains may sink and eventually become part of the sediment on the ocean floor. Over geologic time, the sediment turns into limestone, which is the largest carbon reservoir on Earth.

On land, carbon is stored in soil as organic carbon from the decomposition of living organisms or as inorganic carbon from weathering of terrestrial rock and minerals. Deeper under the ground are fossil fuels such as oil, coal, and natural gas, which are the remains of plants decomposed under anaerobic—oxygen-free—conditions. Fossil fuels take millions of years to form. When humans burn them, carbon is released into the atmosphere as carbon dioxide.

Another way for carbon to enter the atmosphere is by the eruption of volcanoes. Carbon-containing sediments in the ocean floor are taken deep within the Earth in a process called subduction, in which one tectonic plate moves under another. This process forms carbon dioxide, which can be released into the atmosphere by volcanic eruptions or hydrothermal vents.

Human Impacts on the Carbon Cycle

Global demand for Earth's limited fossil fuel reserves has risen since the beginning of the Industrial Revolution. Fossil fuels are considered a nonrenewable resource because they are being used up much faster than they can be produced by geological processes.

When fossil fuels are burned, carbon dioxide — CO_2—is released into the air. Increasing use of fossil fuels has led to elevated levels of atmospheric CO_2. Deforestaon—the cungdown of forests—is also a major contributor to increasing CO_2 levels. Trees and other parts of a forest ecosystem sequester carbon, and much of the carbon is released as CO_2 if the forest is cleared.

Some of the extra CO_2 produced by human activities is taken up by plants or absorbed by the ocean, but these processes don't fully counteract the increase. So, atmospheric CO_2 levels have risen and connue to rise. CO_2 levels naturally rise and fall in cycles over long periods of me, but they are higher now than they have been in the past 400,000 years, as shown in the graph below:

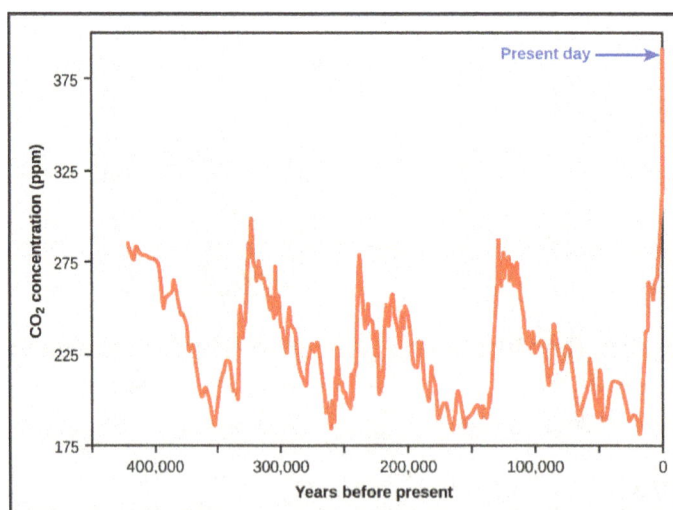

CO_2 is a greenhouse gas. When in the atmosphere, it traps heat and keeps it from radiang into space. Based on extensive evidence, scients think that elevated levels of CO_2 and other greenhouse gases are causing pronounced changes in Earth's climate. Without decisive changes to reduce emissions, Earth's temperature is projected to increase by 1 to 5 C by the year 2100.

Also, while uptake of excess carbon dioxide by the oceans might seem good from a greenhouse gas perspecve, it may not be good at all from the perspecve of sea life. CO_2 dissolved in seawater can react with water molecules to release H^+ ions. So, dissolving more CO_2 in water causes the water to become more acidic. More acidic water can, in turn, reduce CO_3^{2-} concentraons and make it harder for marine organisms to build and maintain their shells of $CaCO_3$. Both increasing temperatures and higher acidity can harm sea life and have been linked to coral bleaching.

A bleached coral appears in the front, while a healthy, unbleached, brown coral appears in the background.

The debate about the future effects of increasing atmospheric carbon on climate change focuses on fossils fuels. However, scientists must take natural processes, such as volcanoes, plant growth, soil carbon levels, and respiration, into account as they model and predict the future impact of this increase.

References

- Biogeochemical-cycles, air-climate-weather: enviroliteracy.org, Retrieved 07 January, 2019

- Oxygen-cycle: britannica.com, Retrieved 08 May, 2019

- Oxygen-cycle-environment: byjus.com, Retrieved 18 June, 2019

- Nitrogen-cycle, biogeochemical-cycles: enviroliteracy.org, Retrieved 15 August, 2019

- The-nitrogen-cycle, biogeochemical-cycles: khanacademy.org, Retrieved 19 April, 2019

- Phosphorus-cycle-definition-steps-facts: eartheclipse.com, Retrieved 09 February, 2019

- Carbon-cycle: scied.ucar.edu, Retrieved 16 March, 2019

- The-carbon-cycle, biogeochemical-cycles: khanacademy.org, Retrieved 18 June, 2019

Environmental Pollution

The contamination of the physical and biological components of the earth and atmosphere is referred to as environmental pollution. It can be categorized into soil pollution, air pollution and water pollution. The topics elaborated in this chapter will help in gaining a better perspective about these types of environmental pollution.

Environmental pollution has existed for centuries but only started to be significant following the industrial revolution in the 19th century. Pollution occurs when the natural environment cannot destroy an element without creating harm or damage to itself. The elements involved are not produced by nature, and the destroying process can vary from a few days to thousands of years (that is, for instance, the case for radioactive pollutants). In other words, pollution takes place when nature does not know how to decompose an element that has been brought to it in an unnatural way.

Pollution must be taken seriously, as it has a negative effect on natural elements that are an absolute need for life to exist on earth, such as water and air. Indeed, without it, or if they were present on different quantities, animals – including humans – and plants could not survive. We can identify several types of pollution on Earth: air pollution, water pollution and soil pollution.

Causes of Environmental Pollution

Let us first take a look at the causes of environmental pollution:

- Industries: Industries have been polluting our environment especially since the beginning of the industrial revolution, as mentioned above, notably due to the increasing use of fossil fuels. In the 19th century and for a significant part of the 20th century, coal has been use to make machines work faster, replacing human force. Though pollution by industries mainly causes air pollution, soil and water contamination can also occur. This is particularly the case for power-generating industries, such as plants producing electricity (May they be a dam, a nuclear reactor or some other type of plant).

  Also, the transportation of this energy can be harmful to the environment. We can take as an example the transportation of petrol through pipelines; if there is a leak in the pipeline, soil will automatically be polluted. At the same time, if the tanker transporting the petrol from its production plant to the place where it will be consumed leaks or sinks, the water will get contaminated.

- Transportation: Ever since men abandoned animal power to travel, pollution of the environment has become higher and higher. Its levels have only been increasing until now. Similarly to industries, pollution caused by transport can mainly be attributed to fossil fuels. Indeed, humans went from horse carriages to cars, trains (which, before electricity, used to be propelled by coal), and airplanes. As the traffic is increasing every day, pollution follows that evolution.

- Agricultural Activities: Agriculture is mainly responsible for the contamination of water and soil. This is caused by the increased use of pesticides, as well as by the intensive character of its production. Almost all pesticides are made from chemical substances and are meant to keep diseases and threatening animals away from the crops. However, by keeping these forms of life away, harm is almost always made to the surrounding environment as well.

  Furthermore, as agriculture gets more and more intensive to feed the increasing world population, more environments and ecosystems are destroyed to make space for the crops. Some of them, like rapeseed –used to make oil – demand a lot of space for a relatively small output.

- Trading Activities: Trading activities including the production and exchange of goods and services. Concerning goods, pollution can be caused by packaging (which often involves the use of plastic, which is made from fossil fuels) or transport, mainly.

- Residences: Finally, residential areas provide their fair share of pollution as well. First, to be able to build homes, natural environment has to be destroyed in one way or another. Wildlife and plants are driven away and replaced by human constructions. As it requires the work of industries, construction itself is also a source of contamination of the environment. Then, when people settle in, they will produce waste every day, including a part that cannot be processed by the environment without harm yet.

Soil Pollution

Soil Pollution has gradually become a major challenge that we need to overcome for establishing a healthy environment. Weathering of the earth's crusts by different processes leads to the formation of soil that accumulates over the centuries. The soil is the home for a large part of bacterial biodiversity and other microscopic and macroscopic living organisms.

Soil Pollution.

Therefore, it is very important to judiciously use pesticides because it contains lots of different harmful chemicals. Therefore, to improve soil and prevent soil pollution it is important to limit the use of pesticides and herbicides.

Soil pollution refers to anything that causes contamination of soil and degrades the soil quality. It occurs when the pollutants causing the pollution reduce the quality of the soil and convert the soil inhabitable for microorganisms and macro organisms living in the soil.

Soil contamination or soil pollution can occur either because of human activities or because of natural processes. However, mostly it is due to human activities. The soil contamination can occur due to the presence of chemicals such as pesticides, herbicides, ammonia, petroleum hydrocarbons, lead, nitrate, mercury, naphthalene, etc., in an excess amount.

The primary cause of soil pollution is a lack of awareness in general people. Thus, due to many different human activities such as overuse of pesticides the soil will lose its fertility. Moreover, the presence of excess chemicals will increase the alkalinity or acidity of soil thus degrading the soil quality. This will in turn cause soil erosion. This soil erosion refers to soil pollution.

Causes of Soil Pollution

Soil pollution can be natural or due to human activity. However, it mostly boils down to the activities of the human that causes the majority of soil pollution such as heavy industries, or pesticides in agriculture.

Pesticides

Before World War II, the chemical nicotine chemical present in the tobacco plants was used as the pest controlling substance in agricultural practices. However, DDT was found to be extremely useful for malaria control and as pest control of many insects during World War II. Therefore, it was used for controlling many diseases.

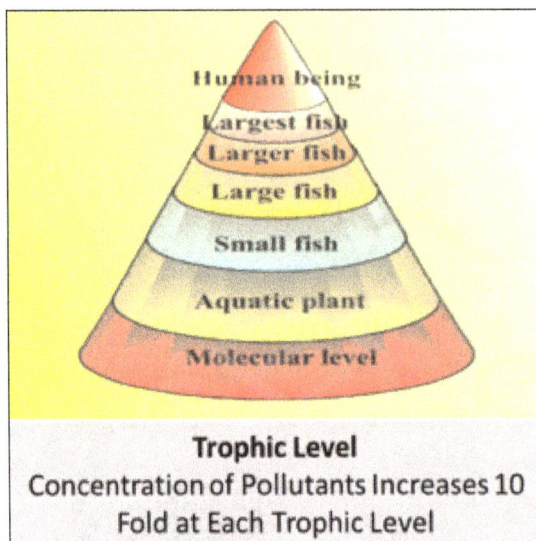

Trophic Level
Concentration of Pollutants Increases 10
Fold at Each Trophic Level

Hence, post-war, people started using it as pest control in agriculture for killing rodents, weeds, insects, etc and avoiding the damages due to these pests. However, everyone gradually the adverse effects of this chemical which led to the ban of this chemical in many parts of the world including India.

Moreover, pests became resistance to DDT due to the chemicals regular use. Hence this led to the introduction of other harmful chemicals such as Aldrin and Dieldrin. Pesticides are synthetic toxic chemicals that definitely kill different types of pests and insects causing damage to agriculture but it has many ecological repercussions.

They are generally insoluble in water and non-biodegradable. Therefore, these chemicals will not gradually decompose and keep on accumulating in the soil. Therefore, the concentration of these chemicals will increase when the transfer of these chemicals take place from lower to higher trophic level via the food chain. Hence, it will cause many metabolic and physiological disorders in humans.

Chlorinated Organic Toxins

The harmful effect of DDT and other chemicals led to the introduction of less persistent organic and more-biodegradable substance such as carbamates and organophosphates. However, these chemicals act as harmful toxins for nerves, hence they are more dangerous to humans. It led to pesticides related to the death of field workers in some agricultural fields.

Herbicides

Slowly, the industries began production of herbicides like sodium arsenite (Na_3AsO_3), sodium chlorate ($NaClO_3$), etc. Herbicides can decompose in a span of few months. However, even they affect the environment and are not environmental friendly. Even though they are not as harmful as organo-chlorides but most of the herbicides are toxic. They are known to cause birth defects.

Furthermore, research suggests that spraying herbicides causes more insect attack and diseases of plants in comparison to manual weeding. One thing to note here is all the above factors occupy just a small portion of the causes. Majority of the causes is related to manufacturing activities in chemical and industrial processes that are released in nature or environment.

Inorganic Fertilizers

Excessive use of inorganic nitrogen fertilizers leads to acidification of soil and contaminate the agricultural soil, also known as agrochemical pollution.

Industrial Pollution

The incorrect way of chemical waste disposal from different types of industries can cause contamination of soil. Human activities like this have led to acidification of soil and contamination due to the disposal of industrial waste, heavy metals, toxic chemicals, dumping oil and fuel, etc.

Inferior Irrigation Practices

Poor irrigation methods increase the soil salinity. Moreover, excess watering, improper maintenance of canals and irrigation channels, lack of crop rotation and intensive farming gradually decreases the quality of soil over time and cause degradation of land.

Solid Waste

Disposal of plastics, cans, and other solid waste falls into the category of soil pollution. Disposal of electrical goods such as batteries causes an adverse effect on the soil due to the presence of harmful chemicals. For instance, lithium present in batteries can cause leaching of soil.

Urban Activities

Lack of proper waste disposal, regular constructions can cause excessive damage to the soil due to lack of proper drainage and surface run-off. These waste disposed of by humans contain chemical waste from residential areas. Moreover leaking of sewerage system can also affect soil quality and cause soil pollution by changing the chemical composition of the soil.

After-effects of Soil Pollution

Soil pollution is not only the problem in India but it is a global problem. It causes harmful effect on the soil and the environment at large. Contamination of soil will decrease the agricultural output of a land. Major soil pollution after effects are:

Inferior Crop Quality

It can decrease the quality of the crop. Regular use of chemical fertilizers, inorganic fertilizers, pesticides will decrease the fertility of the soil at a rapid rate and alter the structure of the soil. This will lead to decrease in soil quality and poor quality of crops. Over the time the soil will become less productive due to the accumulation of toxic chemicals in large quantity.

Harmful Effect on Human Health

It will increase the exposure to toxic and harmful chemicals thus increasing health threats to people living nearby and on the degraded land. Living, working or playing in the contaminated soil can lead to respiratory diseases, skin diseases, and other diseases. Moreover, it can cause other health problems.

Water Sources Contamination

The surface run-off after raining will carry the polluted soil and enter into different water resource. Thus, it can cause underground water contamination thereby causing water pollution. This water after contamination is not fit for human as well as animal use due to the presence of toxic chemicals.

Negative Impact on Ecosystem and Biodiversity

Soil pollution can cause an imbalance of the ecosystem of the soil. The soil is an important habitat and is the house of different type of microorganisms, animals, reptiles, mammals, birds, and insects. Thus, soil pollution can negatively impact the lives of the living organisms and can result in the gradual death of many organisms. It can cause health threats to animals grazing in the contaminated soil or microorganisms residing in the soil.

Therefore, human activities are responsible for the majority of the soil pollution. We as humans buy things that are harmful and not necessary, use agricultural chemicals (fertilizers, pesticides, herbicides, etc.), drop waste here and there. Without being aware we harm our own environment.

Therefore, it is very important to educate people around you the importance of environment if they are not aware. Prevention of soil erosion will help to cease soil pollution. Thus, it is our small steps and activities that can help us to achieve a healthier planet for us. Therefore, it is essential for industries, individuals and businesses to understand the importance of soil and prevent soil pollution and stop the devastation caused to plant and animal life.

Soil Erosion

Soil erosion is, at its core, a natural process. Put simply, it is when topsoil, which is the upper-most layer of the ground, is moved from one spot to another. Why this matters is because topsoil is the part of the land that is highest in organic matter and best suited for farming and other fertile activities, which is why soil erosion can have the greatest impact on farmers and agricultural land. In other words, soil erosion is a naturally occurring and slow process that refers to loss of field's top soil by water and wind or through conversion of natural vegetation to agricultural land.

When farming activities are carried out, the top soil is exposed and is often blown away by wind or washed away by rain. When soil erosion occurs, the movement of the detached topsoil is typically facilitated by either a natural process – such as wind or water movement – or by the impact of man, such as through tilling farmland.

The process of soil erosion is made up of three parts:

- Detachment: This is when the topsoil is actually "detached" from the rest of the ground.

- Movement: This is when the topsoil is relocated to another area.

- Deposition: Where the topsoil ends up after this process.

When it comes to our planet, natural resources are typically affected by two things – either naturally-occurring ones such as weather, or from man-made influence. Soil erosion, or the gradual reduction of topsoil in a geographic area, can be caused by both natural and unnatural processes, but it can also have great effects on inhabitants of an affected area. One of the major concerns regarding soil erosion is that it can permanently affect the land, which can be devastating for farmers or those with agricultural pursuits.

Unfortunately, many people are still uneducated about soil erosion, which is leading to the occurrence in greater amounts around the world. Soil erosion contributes to pollution in adjacent water sources and reduces cropland productivity. Major crops that cause soil erosion include coffee, cotton, tea, tobacco, palm oil, soybean and wheat that can increase soil erosion beyond the soil's ability to maintain itself.

Causes of Soil Erosion

The predominant causes of soil erosion are either related to naturally-occurring events or influenced by the presence of human activity. Some of the principal causes of soil erosion include:

- Rain and rainwater runoff: In a particular heavy rain, soil erosion is common. First of all, the water starts to break down the soil, dispersing the materials it is made of. Typically, rainwater runoff will impact lighter materials like silt, organic matter, and finer sand particles, but in heavy rainfall, this can also include the larger material components as well.

- Farming: When land is worked through crops or other agricultural processes, it reduces the overall structure of the soil, in addition to reducing the levels of organic matter, making it more susceptible to the effects of rain and water. Tilling in particular, because it often breaks up and softens the structure of soil, can be a major contributor to erosion. Farming practices that reduce this activity tend to have far less issues with soil erosion.

- Slope of the land: The physical characteristics of the land can also contribute to soil erosion. For example, land with a high hill slope will perpetuate the process of rainwater or runoff saturation in the area, particularly due to the faster movement of the water down a slope.

- Lack of vegetation: Plants and crops help maintain the structure of soils, reducing the amount of soil erosion. Areas with less naturally-occurring flora may be a hint that the soil is prone to erosion.

- Wind: Wind can be a major factor in reducing soil quality and promotion erosion, particularly if the soil's structure has already been loosened up. However, lighter winds will typically not cause too much damage, if any. The most susceptible soil to this type of erosion is sandy or lighter soil that can easily be transported through the air.

Effects of Soil Erosion

A major problem with soil erosion is that there is no telling how quickly or slowly it will occur. If largely impacted by ongoing weather or climate events, it may be a slow-developing process that is never even noticed. However, a severe weather occurrence or other experience can contribute to rapid-moving erosion, which can cause great harm to the area and its inhabitants.

Some of the greatest effects of soil erosion include:

- Loss of topsoil: Obviously, this is the biggest effect of soil erosion. Because topsoil is so fertile, if it is removed, this can cause serious harm to farmer's crops or the ability to effectively work their land.

- Soil compaction: When soil under the topsoil becomes compacted and stiff, it reduces the ability for water to infiltrate these deeper levels, keeping runoff at greater levels, which increases the risk of more serious erosion.

- Reduced organic and fertile matter: As mentioned, removing topsoil that is heavy with organic matter will reduce the ability for the land to regenerate new flora or crops. When new crops or plants can't be placed successfully in the area, this perpetuates a cycle of reduced levels of organic nutrients.

- Poor drainage: Sometimes too much compaction with sand can lead to an effective crust that seals in the surface layer, making it even harder for water to pass through to deeper layers. In some ways, this can help erosion because of the densely packed soil, but if it perpetuates greater levels of runoff from rainwater or flooding, it can negatively impact the crucial topsoil.

- Issues with plant reproduction: When soil is eroded in an active cropland, wind in particular makes lighter soil properties such as new seeds and seedlings to be buried or destroyed. This, in turn, impacts future crop production.

- Soil acidity levels: When the structure of the soil becomes compromised, and organic matter is greatly reduced, there is a higher chance of increased soil acidity, which will significantly impact the ability for plants and crops to grow.

- Long term erosion: Unfortunately, if an area is prone to erosion or has a history of it, it becomes even harder to protect it in the future. The process has already reduced the soil structure and organic matter of the area, meaning that it will be harder to recover in the long run.

- Water pollution: A major problem with runoff from soils – particularly those used for agricultural processes – is that there is a greater likelihood that sediment and contamination like the use of fertilizer or pesticide. This can have significant damage on fish and water quality.

Solutions for Soil Erosion

When it comes to finding solutions for soil erosion, the most useful techniques found tend to be those that emphasize reinforcing the structure of the soil, and reducing processes that affect it:

- Careful tilling: Because tilling activity breaks up the structure of soil, doing less tilling with fewer passes will preserve more of the crucial topsoil.

- Crop rotation: Plenty of crop rotation is crucial for keeping land happy and healthy. This allows organic matter to build up, making future plantings more fertile.

- Increased structure for plants: Introducing terraces or other means of stabilizing plant life or even the soil around them can help reduce the chance that the soil loosens and erodes. Boosting areas that are prone to erosion with sturdy plant life can be a great way to stave off future effects.

- Water control: For those areas where soil erosion is predominantly caused by water – whether natural or man-made – specialized chutes and runoff pipes can help to direct these water sources away from the susceptible areas, helping stave off excess erosion. Having these filters in particular areas rather than leading to natural bodies of water is a focus to reduce pollution.

- Increased knowledge: A major factor for preventing soil erosion is educating more and more people who work with the land on why it is a concern, and what they can do to help reduce it. This means outreach to farmers in susceptible areas for ways that they can help protect crops from inclement weather, or ways that they can help make sure their soil remains compact without restricting their plant growing activities.

Air Pollution

Air pollution means the presence of chemicals or compounds (called pollutants) in the air which are not naturally occurring, and which lower the quality of air, and are harmful to all living things in the atmosphere. Air pollution is majorly caused due to the release of various chemicals into the atmosphere. Air pollution can be both man-made and naturally occurring.

In our current age of industrialisation and modernisation, the biggest source of air pollution is the burning of fossil fuels. For example when we burn petrol or diesel or coal to run our cars, machines, trains, power plants etc. this releases harmful pollutants into the atmosphere, endangering all living things around.

There are two major types of air pollutants, gaseous compounds and compounds in solid form. There is actually a laundry list of the various pollutants that are affecting our environment, but the few very dangerous ones are:

- Carbon Monoxide.

- Sulphur Oxides.

- Nitrogen Oxides.

- Carbon Dioxide.

- Ammonia.

- Particulate Matter.

- Radioactive Pollutants.

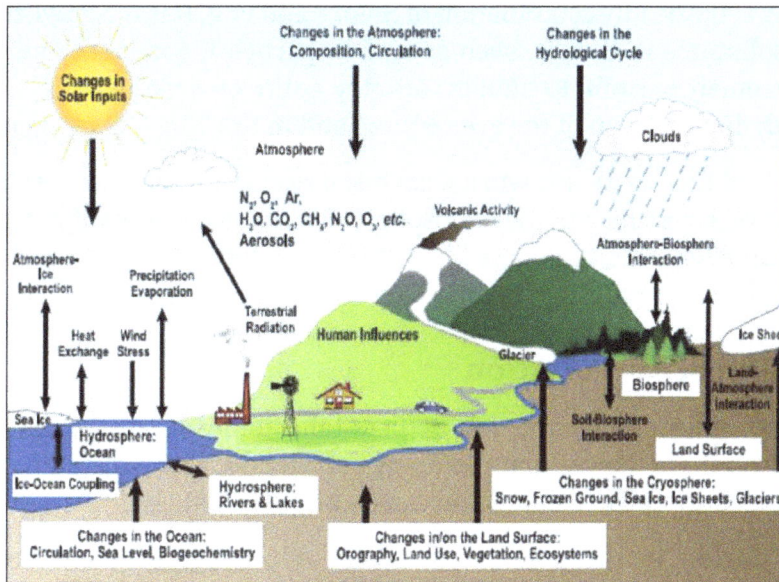

- Water and Water Pollution.

- Mineral riches in the soil.

- Biogeochemical Cycle and its role.

Causes of Air Pollution

Air pollution can be caused by both man-made and natural causes, although the contribution by these natural causes is pretty negligible.

The main culprit of air pollution is the man-made sources of air pollution. The single most harmful source of air pollution is the unchecked burning of fossil fuels by mankind. Fossil fuels (non-renewable sources of energy such as crude oil, petrol, diesel, coal etc.) are used in almost every process of industrialization, manufacturing, transport and energy generation. In rural areas, a major source of pollution is the practice of unchecked crop burning. In moderation, this is actually a useful tool in farming but uncontrolled crop burning causes significant air pollution. Another source of man-made pollution is military resources such as nuclear arsenal and chemical weaponry.

There are a few natural sources of air pollution as well, such as forest fires, volcanic activity and methane discharged from cattle. However, the amount of air pollution caused by natural causes is minuscule when compared to the damage done by man-made causes.

Consequences of Air Pollution

- Ozone Depletion: The Ozone layer is a belt of natural gasses in the stratosphere that envelopes the planet. It protects us from harmful ultraviolet B radiation that the Sun emits. The pollutants that cause air pollution in our atmosphere are seriously damaging the ozone layer. Over the last few decades, many holes have appeared in the layer. The ozone layer above Antartica particularly has seen some very serious damage. When the ozone layer gets depleted the harmful UVB radiation reaches the surface of the earth and causes skin cancer and eye damage among humans.

- Smog: Smog is the deadly combination of Smoke and Fog. It is the greyish fog that is caused when the pollutants in the air, such as carbon particles, condense and mix with the fog. Smog is extremely harmful to humans and the entire environment. It can lead to diseases such as cold, flu, irritation of the eye, asthma and in the long term even lung cancer.

- Acid Rain: Acid Rain is caused when a chemical reaction occurs between air pollutants and water and oxygen very high in the atmosphere. Pollutants such as sulphur dioxide and nitrogen oxide are water soluble and mix with the water in the atmosphere and precipitate as acid rain. The acidic nature of this rain causes severe damage to the plants, animals and the soil.

Strategies to Control Air Pollution

- Use of Energy efficient appliances: Whether at the domestic level or at the industrial level, we must push for appliances that use energy efficiently, which result in complete combustion of fuel, as incomplete combustion causes air pollution.

- Shifting industries: Another possible solution to reduce the harmful effects of air pollution is to shift the manufacturing plants, factories and industries to remote areas with a low level of population. By doing so you can ensure that pollution in the urban areas decreases over time.

- Using Modern Techniques: With technology making great advancements, there are now technologies available that can help reduce the release of pollutants in the air. Air filters, scrubbers, precipitators are just a few examples.

- Shifting to Natural Gasses: Instead of using and exhausting fossil fuels, shifting to greener options is a no-brainer. For example, using CNG (compressed natural gas) instead of petrol or diesel is a great option.

Air Pollutants

Any substance in air that could, in high enough concentration, harm animals, humans, vegetation, and/or materials. Such pollutants may be present as solid particles, liquid droplets, or gases. Air pollutants fall into two main groups: (1) those emitted from identifiable sources and, (2) those formed in the air by interaction between other pollutants. Over one hundred air pollutants have been identified, which include halogen compounds, nitrogen compounds, oxygen compounds, radioactive compounds, sulphur (sulfur) compounds, and volatile organic chemicals (VOC).

Primary and Secondary Air Pollutants

Air Pollutants can be solid particles, liquid droplets, or gases. They can be natural or manmade. The pollutants have been classified into primary and secondary categories.

- The primary pollutants are "directly" emitted from the processes such as fossil fuel consumption, Volcanic eruption and factories. The major primary pollutants are Oxides of Sulphur, Oxides of Nitrogen, Oxides of Carbon, Particulate Matter, Methane, Ammonia, Chlorofluorocarbons, Toxic metals etc.

- The secondary pollutants are not emitted directly. The secondary pollutants form when the primary pollutants react with themselves or other components of the atmosphere. Most important secondary level Air Pollutants are Ground Level Ozone, Smog and POPs (Persistent Organic Pollutants).

Primary Air Pollutants

Oxides of Sulphur

Sulphur Oxides are generally a product of the Volcanoes, Industrial processes, Coal and petroleum, because most of them have Sulphur as a component. The Sulphur Dioxide in presence of a catalyst such as NO_2 causes Acid Rain, because of the formation of Sulphuric Acid.

Oxides of Nitrogen

Most of the Nitrogen Oxides are produced due to high temperature combustion. In the cities the brown haze dome above the cities is mostly because of the Nitrogen Oxides. The most important toxic gas is Nitrogen dioxide which is brown, with sharp odour.

Oxides of Carbon

Carbon Monoxide, which is colourless, odourless and non irritating but very poisonous gas is the product of incomplete combustion of the natural gas, coal or wood. The vehicle exhaust is the major source of CO.

The Carbon Dioxide is associated with the Ocean Acidification and is emitted from combustion, factories and respiration of living organisms.

Then, we have primary pollutants such as Volatile Organic Compounds or VOCs which are methane (CH_4) and non-methane (NMVOCs).

- Methane is a GHG which contributes to Global Warming.

- The NMVOCs include the aromatic compounds such as Benzene, Toluene, Xylene which are proved or suspected carcinogens.

- Another dangerous compound is the 1,3-butadiene, often associated with industrial uses.

Particulate Matter

The particulate matters are the fine particles which may be either solid or liquid, suspended in a gas. They are different from the Aerosols. Aerosols are particle and gas referred together. The aerosols which are created by the Human activities are anthropogenic aerosols. They account for around 10% of the total aerosols in the atmosphere.

Other Primary Pollutants

Another category of the primary air pollutants is toxic metals such as Cadmium, Lead and Copper, which are products of the Industrial processes. The Chlorofluorocarbons (CFCs) are proved to be harmful to the ozone layer emitted from products currently banned from use. In agriculture process, Ammonia is emitted which has characteristic pungent odor. It is a precursor to foodstuffs and fertilizers. Ammonia is also a building block for the synthesis of many pharmaceuticals.

Secondary Air Pollutants

Ground Level Ozone

The most important secondary pollutant is the Ground Level Ozone or Tropospheric Ozone. Emissions from industrial facilities and electric utilities, motor vehicle exhaust, gasoline vapours, and chemical solvents are some of the major sources of No_x and VOC. Ground Level Ozone forms due to reactions of the NO_x, Carbon Monoxide and VOCs in presence of sunlight.

Smog

Another most important secondary pollutant is the Smog, which has made up of Smoke and Fog. Traditionally, the smog has resulted from large amounts of coal burning in an area caused by a mixture of smoke and sulphur dioxide. Now-a-days, the Vehicle emissions and Industrial emissions that are acted on in the atmosphere by ultraviolet light from the sun to form secondary pollutants that also combine with the primary emissions to form photochemical smog.

Difference between Primary Pollutants and Secondary Pollutants

A primary pollutant is an air pollutant emitted from a source directly into the atmosphere.

A secondary pollutant is an air pollutant formed in the atmosphere as a result of the chemical or the physical interactions between the primary pollutants themselves or between the primary pollutants and other atmospheric components.

Examples of primary pollutants include sulfur dioxide (SO_2), carbon monoxide (CO), nitrogen oxides (NO_x), and particulate matter (PM).

Examples of secondary pollutants include photochemical oxidants (ozone, nitrogen dioxide, sulfur trioxide) and secondary particulate matter.

Pollution Effect

Primary pollutants are considered chemical reactants, being involved in the chemical reactions that result in the formation of secondary pollutants. Therefore, their pollution effect can either be direct like the impact of sulfur dioxide on the human respiratory system or indirect when sulfur dioxide interacts with water in the atmosphere to form acid rain, with serious consequences on the ecosystem.

On the other hand, secondary pollutants are chemical products inferring that they might be more stable and inert with a limited polluting effect. While this might be true to some extent, it is not the case of ozone where photoactivation is involved, making the chemical process highly reactive.

Pollution Control

The atmospheric concentration of primary pollutants can be controlled in a direct way through the reduction of anthropogenic emissions.

On the contrary, controlling the secondary pollutants is a much more complicated process: the chemical reactions involved in their formation must be understood and interrupted.

Primary Pollutants	Secondary Pollutants
Air pollutant emitted directly from a source into the atmosphere.	Air pollutant formed in the atmosphere as a result of chemical or physical interactions between the primary pollutants themselves or between the primary pollutants and other atmospheric components.
Sulfure dioxide (SO_2), carbon monoxide (CO), nitrogen oxide (NO_x), and particulate matter (PM).	Photochemical oxidants (ozone nitrogen dioxide, sulfur trioxide) and secondary particulate matter.
Chemical reactants characterized with a direct pollution effect through the formation of secondary pollutants.	Chemical products, highly reactive when photoactivation is involved in the chemical process of their formation.
Direct control through the reduction of anthropogenic emissions.	Complicated control process: understanding and interrupting the chemical reactions leading to their generation.

Primary and secondary pollutants are two groups of air polluting substances differing mainly by their sources of emission or generation.

Primary pollutants are emitted from natural or anthropogenic sources directly into the atmosphere, while secondary pollutants result from the chemical reactions or the physical interactions between the primary pollutants themselves or between the primary pollutants and other atmospheric components.

Examples of primary pollutants are sulfur dioxide, nitrogen oxides, carbon monoxide, and primary particulate matter. Examples of secondary pollutants are photochemical oxidants like ozone and secondary particulate matter.

Identifying air pollutants and studying the difference between primary and secondary pollutants are important for the control and prevention of atmospheric pollution, especially through the reduction of anthropogenic sources.

Water Pollution

Water is a necessary component for the sustenance of life. Even after knowing how essential water is for the sustenance of life, we still take water for granted and waste it time and again. Mostly, it is the human activities that have led to water pollution and compromised with the quality of water.

Water pollution reaches the surface or grounds via different paths. The places of pollution or source that can be easily identified easily called point source. In the case of the point source such as industrial and municipal discharge pipes, the pollutant enters via these mediums in the water source.

The identification of certain pollution sources is difficult such as agricultural run-offs, stormwater drainage, acid rain, etc. These sources of pollution are non-point sources. It is difficult to recognize them because it is possible to have many agricultural runoffs like farms, animals or crop lands from where water can enter into the water source. Below is a list of major water pollutants and the different sources from it originates to finally enter the water bodies.

Category	Examples	Sources
1. Affecting health		
Infectious agents	Bacteria, viruses and parasites	Sewage, human and animal excreta
Organic chemicals	Pesticides, plastics, detergents, oil	Agricultural, industrial and domestic wastes
Inorganic chemicals	Acids, caustics, salts, metals	Industrial and domestic effluents
Radioactive materials	Uranium, thorium, radon, etc.	Mining, power plants, natural sources
2. Affecting ecosystem		
Plant nutrients	Nitrates, phosphates, etc	Chemical fertilisers, sewage, manure
Sediments	Silt, soil	Soil erosion
Thermal	Heat	Industries, power plants
Oxygen demanding	Agricultural wastes, manure	Sewage, agricultural runoff

Causes of Water Pollution

Contamination of Water by Pathogens

Disease-causing agents or pathogens is one of the major water pollutants responsible for contamination of water and causes of water pollution. Domestic sewage and animal excreta contain different types of pathogens such as bacteria and other microorganisms. For instance, human excreta contain bacteria like Streptococcus faecalis and E. coli responsible for gastrointestinal diseases.

Water Pollution due to Organic Wastes

Another very important pollutant is organic matter like leaves, trash, grass, dried leaves, etc. The runoff of the organic matter leads to the pollution of water. Excessive growth of phytoplankton in water can also be among the causes of water pollution. However, these wastes are biodegradable (waste that can decompose).

The large bacterial population present in the water can decompose the organic matter. These bacteria consume the oxygen present in the water. However, the amount of oxygen present in the water depends upon the holding capacity of the water.

The oxygen present in water is the concentration of dissolved oxygen (DO). The DO concentration in cold water is up to 10 parts per million or 10 ppm. This amount is very small in comparison to oxygen in the air which is around 200,000 ppm. The DO concentration in water is a very important factor in terms of aquatic life.

The inhibition of life can take place if the DO concentration is lower than 6 ppm. Photosynthesis of aquatic plants releases oxygen or the atmospheric oxygen reaching the water is the source of oxygen in any water body. Thus, during the night there is a decrease in the concentration of oxygen in the water because photosynthesis is not possible and the aquatic plants constantly consume oxygen for the respiration process.

However, if organic matter enters the water bodies, the microorganisms present in the organic matter utilize the oxygen available in water. Therefore, this will, in turn, cause the oxygen-dependent aquatic life to slowly diminish and die due to lack of oxygen in the water. Hence, anaerobic bacteria (bacteria that do not require oxygen for their survival) will gradually start decomposing the organic waste and produce foul-smelling chemicals harmful to human health.

Biological Oxygen Demand

In most cases, aerobic bacteria (bacteria that do require oxygen for their survival) utilizes the DO to decompose the organic wastes thereby depleting the DO concentration. Therefore, a particular amount of oxygen is needed by bacteria to decompose the organic matter present in a particular volume of the water sample.

This is known as BOD or Biochemical Oxygen Demand. Therefore, the amount of BOD indicates the amount of organic matter present in a water sample. BOD indicates the amount of oxygen which can easily break down the organic material present in a water body. The BOD value of clean water is 5ppm. The BOD value of highly polluted water can reach up to 17 ppm and sometimes more.

Chemical Pollutants

All of us have read that water is an excellent solvent. Therefore it can dissolve even harmful heavy metals like mercury, cadmium, nickel, etc. These types of heavy metals fall into the category of important types of pollutants because they are very harmful to human body.

Moreover, human beings are not capable of excreting these heavy metals. It will not show its effect is present in small quantity. However, over the time the concentration of the metals will rise in the human body due to gradual decomposition and can possibly cross the tolerance limit.

Heavy metals if cross the tolerance limit can cause damage in different parts of human body including kidney, liver, central nervous systems, etc. Moreover, another type of chemical pollutants includes acids and salts. An acid such as sulphuric acid can be found in mine drainage and salts can be found in many sources such as sodium and calcium chloride to melt snow and ice in colder regions.

Another set of chemical pollutants responsible for water pollution are organic chemicals. For instance, oil spills (petroleum products) pollute the oceans. Similarly, pesticides are a class of organic substance that can one of the causes of water pollution when they drift down as run-off from lands or in the form of sprays.

Additionally, industrial chemicals such as PCBs or polychlorinated biphenyls are a class of very harmful class water pollutant. Recent studies indicate they are carcinogenic (cancer-causing). PCB is present in many regular household and industrial chemicals like cleansing solvent, detergents, and fertilizers. However, nowadays biodegradable detergents are used.

Eutrophication

The lack of oxygen can be responsible for destroying the aquatic life of any waterbody. Chemical pollutants like fertilizer contain phosphates in the form of additives. Therefore, phosphate addition in water can lead to rapid growth of algae. Hence, gradually the algae growth can cover entire water surface which can lead to the reduction in oxygen concentration.

Therefore, lack of oxygen can cause death and decay of aquatic animals thereby creating anaerobic conditions in the waterbody. Hence, the water body will become bloom-infested and can inhibit the growth of other living organisms. This entire process is eutrophication.

Therefore, eutrophication refers to a process where water bodies support dense plant population due to nutrient enrichment, which can also cause animal death by depriving the animal life of oxygen thereby resulting in gradual loss of biodiversity.

International Standards of Drinking Water

Fluoride

It is important to regularly check and maintain the fluoride ion concentration in drinking water within the permissible limit. Lack of fluoride in drinking water can lead to tooth decay. It is important to add soluble fluoride in drinking water to make the concentration up to 1 ppm. However, the concentration should not cross 2 ppm or it can lead to brown mottling of teeth. Moreover, excessive fluoride can cause harmful effect to teeth and bones.

Lead

Lead pipes are used for water transportation and contaminate the water in the process. The maximum permissible limit for the concentration of lead in drinking water is 50 parts per billion. High concentration of lead in drinking water can cause problems associated with liver, kidney, reproductive system, etc.

Sulfates and Nitrates

Sulfate concentration greater than 500 ppm in drinking water can cause the laxative effect. Nitrate concentration in drinking water should not exceed 50 ppm. High nitrate concentration in drinking water leads to harmful effects such as methemoglobinemia or blue baby syndrome.

Other Metals

Many other types of metals can be present in drinking water. However, the permissible limit for each metal may vary from each other. Below table represents the permissible concentration of few metals in drinking water.

Heavy metal	Permissible limit				
	WHO	USEPA	ISI	CPCB	ICMR
Iron (mg/l)	0.1	—	0.3	1.0	1.0
Cupper (mg/l)	1.0	1.3	0.05	1.5	1.5
Mercury (mg/l)	0.001	0.002	0.001	No relaxation	0.001
Cadmium (mg/l)	0.005	0.005	0.01	No relaxation	0.01
Arsenic (mg/l)	0.05	0.05	0.05	No relaxation	0.05
Lead (mg/l)	0.05	—	0.10	No relaxation	0.05
Zinc (mg/l)	5.0	—	5.0	15.0	0.10
Chromium (mg/l)	0.1	—	0.05	No relaxation	—

References

- Causes-and-effects-of-environmental-pollution: conserve-energy-future.com, Retrieved 04 June, 2019

- Soil-pollution, environmental-chemistry: toppr.com, Retrieved 16 July, 2019

- Causes-effects-solutions-of-soil-erosion: conserve-energy-future.com, Retrieved 19 April, 2019

- Air-and-air-pollution, natural-resources: toppr.com, Retrieved 18 March, 2019

- Air-pollutant: businessdictionary.com, Retrieved 12 July, 2019

- Primary-secondary-air-pollutants: gktoday.in, Retrieved 17 April, 2019

- Difference-between-primary-pollutants-and-secondary-pollutants: differencebetween.net, Retrieved 13 June, 2019

- Water-pollution, environmental-chemistry: toppr.com, Retrieved 16 May, 2019

Environmental Problems

Environmental problems are the unwanted changes that are caused by natural and man-made activities in the environment. It includes acid rain, ozone depletion, photochemical smog, global warming, etc. This chapter closely examines these environmental problems to provide an extensive understanding of the subject.

Ozone Depletion

Ozone depletion is gradual thinning of Earth's ozone layer in the upper atmosphere caused by the release of chemical compounds containing gaseous chlorine or bromine from industry and other human activities. The thinning is most pronounced in the polar regions, especially over Antarctica. Ozone depletion is a major environmental problem because it increases the amount of ultraviolet (UV) radiation that reaches Earth's surface, which increases the rate of skin cancer, eye cataracts, and genetic and immune system damage. The Montreal Protocol, ratified in 1987, was the first of several comprehensive international agreements enacted to halt the production and use of ozone-depleting chemicals. As a result of continued international cooperation on this issue, the ozone layer is expected to recover over time.

Ozone Depleting Substances

Ozone depleting substances are man-made gases that destroy ozone once they reach the ozone layer. The ozone layer sits in the upper atmosphere and reduces the amount of harmful ultra violet radiation that reaches Earth from the sun. Ultraviolet radiation can have detrimental effects on both humans and the environment such as inducing skin cancer and cataracts, distorting plant growth and damaging the marine environment.

Ozone depleting substances include:

- Chlorofluorocarbons (CFCs);
- Hydrochlorofluorocarbons (HCFCs);
- Hydrobromoflurocarbons (HBFCs);
- Halons;
- Methyl bromide;
- Carbon tetrachloride;
- Methyl chloroform.

They have been used as:

- Refrigerants in commercial, home and vehicle air conditioners and refrigerators;
- Foam blowing agents;
- Components in electrical equipment;
- Industrial solvents;
- Solvents for cleaning (including dry cleaning);
- Aerosol spray propellants;
- Fumigants.

Chlorofluorocarbons

Chlorofluorcarbons (CFCs) are a group of chemical substances that contain one, two or three carbon atoms and at least one atom each of chlorine and fluorine. CFCs were first developed in the 1920s to replace sulphur dioxide as a coolant gas. In the 1930s they began to replace ammonia for cooling uses. Their non-toxicity, non-flammability, stability and heat-absorption effectiveness earned them early respect as wonder chemicals of the 20th Century. In the late 1940s they began to be used as the propellant in aerosols. This use hit its peak in the late 1970s, when CFC was identified as an ODS and aerosols became the main target of public action. CFCs were also used in the production of packaging, insulation and other foams. In the 1980s they were widely used as coolants in refrigerators and air conditioners, as solvents in degreasers and cleaners and to dilute sterilant gas mixtures, and as blowing agents in the production of foams.

The manufacture and use of CFCs in industry has been severely curtailed following the Montreal Protocol. They have been almost completely phased-out in developed countries, with remaining uses limited to metered-dose inhalers. Starting in 2010, developing countries need to completely phase out CFCs.

Chemical Name	Lifetime (years)	Formula	ODP	GWP (100 yr)
CFC-11 Trichlorofluoromethane	45.0	CCl_3F	1.0	4600
CFC-12 Diclorodifluoromethane (R-12)	100.0	CCl_2F_2	1.0	10600
CFC-113 Trichlorotrifluoroethane	85.0	$C_2Cl_2F_3$	0.8	6000
CFC-114 Dichlorotetrafluoroethane	300.0	$C_2Cl_2F_4$	1.0	9800
CFC-115 Chloropentafluoroethane	1700.0	$CClF_2\text{-}CF_3$	0.6	7200

Hydrochlorofluorocarbons

Hydrochlorofluorocarbons (HCFCs) are a group of chemical substances that contain one, two or three carbon atoms and at least one atom each of hydrogen, chlorine and fluorine. The hydrogen makes them less stable and therefore less damaging to the ozone layer. Most HCFCs have been developed for use as transitional chemicals to replace the more damaging ozone-depleting substances, mainly CFCs. HCFCs have only two to five per cent of the ozone-depleting potential of CFCs, which makes them a good temporary replacement for applications where alternatives completely safe for the ozone layer are not yet available.

HCFCs are mainly used for foam blowing, refrigeration and air conditioning, solvent cleaning and to a lesser extent, aerosols and fire protection. At the 19th Meeting of the Montreal Protocol, it was agreed to accelerate the phase-out of HCFCs in both developed and developing countries. Developed countries agreed to phase-out the production and consumption of HCFCs by 75% in 2010, by 90% in 2015, and complete this accelerated phase-out by 2020, while allowing for the continued use of 0.5% for servicing until 2030.

Chemical Name	Lifetime (years)	Formula	ODP	GWP (100 yr)
HCFC-22 Chlorodifluoromethane	11.8	$CHClF_2$	0.055	1700
HCFC-123 2,2-dichloro-1,1,1-trifluoroethane	1.4	CF_3-CHCl_2	0.02	-
HCFC-124 2-chloro-1,1,1,2-tetrafluoroethane	6,1	$CF_3-CHClF$	0,022	620
HCFC-141b 1,1-dichloro-1-fluoroethane	9.2	CCl_2-CH_3	0.11	700
HCFC-142b 1-chloro-1,1-difluoroethane	18.5	$CClF_2CH_3$	0.065	2400
HCFC-225ca 1,1-dichloro-2,2,3,3,3-pentafluoropropane	2.1	$CHCl_2-CF_2-CF_3$	0.025	-
HCFC-225cb 1,3-dichloro-1,1,2,2,3-pentafluoropropane	6.2	$CClF_2-CF_2-CHClF$	0.033	-

Halons

Halons are a group of chemical compounds containing bromine, chlorine, fluorine or carbon in its structure. Halons are very effective for extinguishing fires. They do not leave a solid residue and are not a threat to people when used at the recommended concentrations. The combined characteristics of halons make them suitable for all types of fire-extinguishing equipment, ranging from industrial/commercial total-flooding equipment to the hand-held fire extinguishers popular for office and home use.

Although halons do not present a direct hazard to people, they have a very high ozone-depleting potential (ODP) because they contain bromine. In fact, halon 1301 - which is largely used in total-flooding systems - has the highest ODP of all known ozone-depleting substances (10 times greater than that of CFC-11).

Developed countries no longer produce new halons but use from stockpiles for airplanes and military applications is still allowed and continues. Developing countries must completely phase-out halons by 2010.

Chemical Name	Lifetime (years)	Formula	ODP	GWP (100 yr)
Halon 1211 Bromochlorodifluoromethane	11.0	$CBrClF_2$	3.0	1300
Halon 1301 Bromotrifluoromethane (R-12B1)	65.0	$CBrF_3$	10.0	6900
Halon 2402 1,2-dibromotetrafluoroethane (R114B2)	-	$CBrF_2-CBrF_2$	6.0	-

Bromochloromethane

Bromochloromethane is a chemical substance that contains chlorine, bromine, carbon and hydrogen. It is also known as Halon 1011 or Chlorobromomethane. Total elimination of consumption and production of bromochloroethane has been in place since January 1st, 2002.

Chemical Name	Lifetime (years)	Formula	ODP	GWP (100 yr)
Bromochloromethane Halon 1011	-	CH_2BrCl	0.12	-

Carbon Tetrachloride

Carbon tetrachloride is a chemical substance consisting of one carbon atom and four chlorine atoms. It was widely used as a raw material (feedstock) in the production of CFCs. Carbon tetrachloride is also used in the manufacture of other chemicals that do not deplete the ozone layer. Smaller quantities of carbon tetrachloride were used in fire extinguishers, as a dry cleaning agent, and as an ingredient in pesticides, pharmaceuticals, paints and solvents.

Total elimination of consumption and production of carbon tetrachloride has been in place since January 1, 1996, with the exception of feedstock for manufacturing substances that do not deplete the ozone layer - this use not being controlled by the Montreal Protocol. Developing countries are due to achieve the same phase-out in 2010.

Chemical Name	Lifetime (Years)	Formula	ODP	GWP (100 yr)
CTC Carbon tetrachloride Tetrachloromethane	35.0	CCl_4	1.1	1800

Methyl Bromide

Methyl bromide is a chemical substance containing bromine, hydrogen and carbon and is an important contributor to ozone depletion. This substance has a wide variety of agricultural uses as a fumigant. It is used for soil fumigation and fumigation of some food production facilities. Methyl bromide is also used in quarantine applications and pre-shipment applications for commodities in trade. This use of methyl bromide is not controlled under the Montreal Protocol.

Developed countries were to achieve the phase-out of agricultural uses by 2005. However, alternatives are either not well known, developed, accepted or allowed by regulatory authorities. Therefore, there are still a number of critical use exemptions under the Protocol, meaning that significant volumes of methyl bromide continue to be used. Developing countries are scheduled to phase-out methyl bromide by 2015.

Chemical Name	Lifetime (years)	Formula	ODP	GWP (100 yr)
MBr Methyl Bromide	0.7	CH_3Br	0.6	-

Methyl Chloroform

Methyl chloroform is a chemical compound consisting of carbon, hydrogen, and chlorine. It was popular because of its versatility and efficiency as a solvent in cleaners, degreasers and adhesives.

It first appeared as a substitute for carbon tetrachloride in the mid-1950s, and by the 1980s was widely used by the electronics and equipment manufacturing industries. Under the Montreal Protocol, its use has been phased-out in developed countries since January 1, 1996 and developing countries have until 2015 to do the same.

Chemical Name	Lifetime (years)	Formula	ODP	GWP (100 yr)
Methyl Chloroform 1,1,1-tri-chloroethane	4.8	$C_2H_3Cl_2$	0.1	140

Hydrobromofluorocarbons

Hydrobromofluorocarbons are a group of chemical substances that contain one, two or three carbon atoms and at least one atom each of hydrogen, bromine and fluorine. They were banned in 1996 in all countries.

Chemical Name	Lifetime (years)	Formula	ODP	GWP (100 yr)
Hydrobromofluorocarbons HBFC	-	C, H, BR, F	0.02 to 7.5	-

Ozone Depleting Potential

The ozone depletion potential (ODP) of a chemical compound is the relative amount of degradation to the ozone layer it can cause, with trichlorofluoromethane (R-11 or CFC-11) being fixed at an ODP of 1.0. Chlorodifluoromethane (R-22), for example, has an ODP of 0.05. CFC 11, or R-11 has the maximum potential amongst chlorocarbons because of the presence of three chlorine atoms in the molecule.

The first proposal of ODP came from Wuebbles in 1983. It was defined as a measure of destructive effects of a substance compared to a reference substance.

Precisely, ODP of a given substance is defined as the ratio of global loss of ozone due to the given substance to the global loss of ozone due to CFC-11 of the same mass.

ODP can be estimated from the molecular structure of a given substance. Chlorofluorocarbons have ODPs roughly equal to 1. Brominated substances have usually higher ODPs in range 5–15, because of more aggressive bromine reaction with ozone. Hydrochlorofluorocarbons have ODPs mostly in range 0.005 - 0.2 due to the presence of the hydrogen which causes them to react readily in the troposphere, therefore reducing their chance to reach the stratosphere where the ozone layer is present. Hydrofluorocarbons (HFC) have no chlorine content, so their ODP is essentially zero.

ODP is often used in conjunction with a compound's global warming potential (GWP) as a measure of how environmentally detrimental it can be. GWP represents the potential of a substance to contribute to global warming.

In a broad sense, haloalkanes that contain no hydrogen are stable in the troposphere and decompose only in the stratosphere. Those compounds that contain hydrogen also react with OH radicals and can therefore be decomposed in the troposphere, as well. The ozone depletion potential increases with the heavier halogens since the C-X bond strength is lower.

Table: Ozone depleting potential of common compounds.

Compound	R No.	ODP
Trichlorofluoromethane (CCl_3F)	R-11	1
1,1,1,2-Tetrafluoroethane (CF_3-CH_2F)	R-134a	0.000015
Chlorodifluoromethane ($CClF_2$-H)	R-22	0.05
Chlorotrifluoromethane ($CClF_2$-F)	R-13	1
Dichlorodifluoromethane ($CClF_2$-Cl)	R-12	1.00
Bromochlorodifluoromethane ($CClF_2$-Br)		7.9
Carbon tetrachloride (CCl_4)		0.82
Nitrous oxide (N_2O)		0.017
Alkanes (Propane, Isobutane, etc)		0
Ammonia (NH_3)	R-717	0
Carbon dioxide (CO_2)	R-744	0
Nitrogen (N_2)	R-728	0

Greenhouse Effect and Global Warming

The surface of the earth is heating up at a rapid pace. Till date sixteen out of total seventeen warmest years have been recorded after the year 2000. Melting of glaciers, rising of sea levels, decrease in the forests, danger, and scrambling of wildlife.

Global Warming

Scientist interchanges the term global warming with climate change. Although temperature rises in certain areas but the ocean current move the heat around in different parts of the globe. As a result of which some regions experience cooler climate and some experience warmer weathers. Therefore, the climate changes vary in different regions. Hence, the term climate change is generally used by the scientists instead of global warming.

Climatic researchers of WHO suggest global warming is the reason for around 150,000 deaths every year globally, according to an article published in Scientific American in 2009. Global warming is responsible for the rise in temperature. Data from NOAA and NASA indicates that the year 2016 was the hottest year since the inception of climate record keeping from 1895. There has been an overall increase of 0.8 of over the stretch of 100 years.

Earth surface has the ability to absorb almost 75% of the entire solar energy. This is responsible for the increase in temperature. However, part of the energy is given back to the atmosphere in the form of radiation. There are certain gases that have the ability to trap heat and restrict the heat from escaping from the atmosphere.

These gases are called greenhouse gases. Example of greenhouses gases includes ozone, methane, carbon dioxide, chlorofluorocarbon (CFC), etc. These gases absorb heat and add to the process of heating of atmosphere thereby raising the temperature. This phenomenon is global warming.

Greenhouse Effect and its Use

What is greenhouse effect? Was greenhouse effect always a bad news? The answer to the second question is no. It is useful in certain cases. There are certain places where the temperature is quite low. In those places greenhouse room, glass surrounded room helps to grow vegetables, fruits, and flowers. The energy from the sun enters the room.

The room is made up of glass and it helps to retain the heat inside the room. Therefore, the solar energy helps in warming the soil and aiding in the plant growth and vegetation. However, the temperature of the outside will always be lower than the temperate inside the greenhouse. Moreover, the soil and plant release infrared radiation. Thus, the glass reflects part of the energy and absorbs part of it.

Similarly, greenhouse effect helps in the sustenance of life and temperature. The atmosphere of the earth has a blanket of air which surrounds us just like the greenhouse room. The atmosphere has the ability to restrict the heat and trap it around the earth. This phenomenon is a natural greenhouse effect. Thus, it helps in keeping earth warm thereby maintaining temperature and sustenance of life.

The major gas that contributes to absorption of heat is carbon dioxide. Other greenhouse gases are methane, nitrous oxides, CFCs, and ozone. Initially, the production of gases was within permissible quantity. Therefore, it was useful in maintaining the natural balance and life sustenance. However, the growing anthropogenic or human activities are constantly increasing the gases beyond the permissible limit.

Global warming is responsible for melting of glaciers, the rise in sea level, flooding of the coastal regions and an overall increase in the temperature. The rise in temperature is also the reason behind increase incidence of diseases such as yellow fever, malaria, dengue among others. This is Greenhouse Effect.

Example of greenhouse farming.

Greenhouse Effect on a Larger Scale causes Global Warming

We all know that earth is the only planet that can support and sustain life, rest of the planets have extremely cold or hot temperature. The presence of the thin atmospheric layer makes it possible to support life. The thin layer of the atmosphere contains greenhouse gases like CO_2, water vapour, methane gas, nitrous oxide, etc that has the ability to absorb sun rays and warm up the planet.

This prevents the earth from becoming too cold. The absence of greenhouse effect will make the Earth's average temperature to be nearly 60° Fahrenheit, an extremely cold planet. Hence, life would be impossible without the greenhouse effect. However, the concentration of greenhouse gas is on the rise since the 18th century.

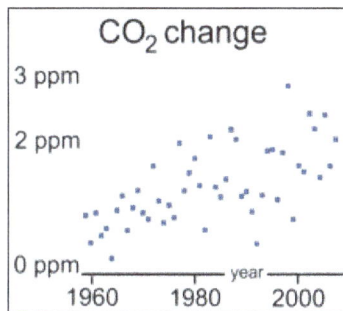

Gradual increase in carbon dioxide concentration every year.

At present, the carbon dioxide level is have increased almost to 40%. The levels of greenhouse gases are at its all-time high than the previous 650,000 years and are continuously rising. This is because the modern life and changes in the lifestyle of humans have led to the constant release of heat-trapping gases or greenhouse gases. This is constantly warming up the atmosphere.

Therefore, this has started increasing the greenhouse gas to an unnatural level and significantly increasing the greenhouse effect. Thus, it is disturbing the environment and is altering the earth's temperature. This phenomenon refers to as Global Warming.

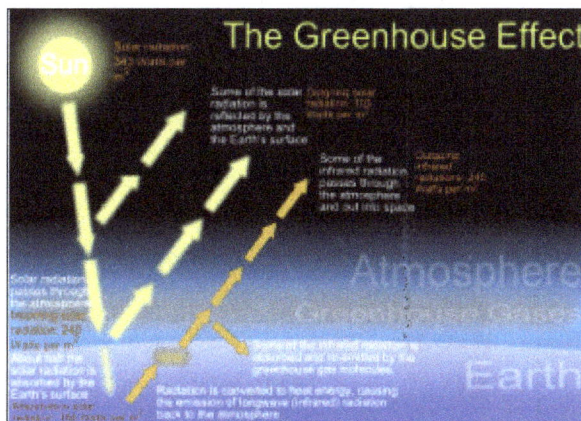

Greenhouse Effect.

It is essential to balance the incoming solar energy by radiating part of energy back to space. We know planet Earth is much colder in comparison to the sun. There are particular chemical compounds or gases present in the atmosphere that allow passing of shortwave radiation (solar energy). This will in turn heat the water bodies and lands.

However, earth after warming up will radiate the energy at much longer wavelengths (infrared light). Land and ocean emit this radiation and the atmosphere along with the clouds absorbs the radiation. Hence, the radiation will again reradiate back to the Earth due to the presence of the greenhouse gases such as carbon dioxide, methane, CFC, nitrous oxide, etc. This is known as greenhouse effect.

Greenhouse Effect: A Rapidly Rising Problem

The greenhouse effect is heating up the Earth's atmosphere. Greenhouse gases are a rapidly rising problem because it is responsible for climatic change at a faster rate. It is becoming difficult for some living organisms to keep pace with this climatic change and unable to adopt the change. This is creating new and unique challenges to all life forms.

Therefore, the amount of the heat the greenhouse gases are trapping is increasing with the gradual increase in the concentration of greenhouse gases present in the atmosphere. Hence greenhouse effect is rapidly raising the problem. Thus, we need to decrease the release of these gases to decrease their quantity in the atmosphere.

Greenhouse Gases

Greenhouse gases include carbon dioxide, methane, and nitrous oxide. Industrial greenhouse gases include hydrofluorocarbons, sulfur hexafluoride, and perfluorocarbons.

Methane

The production and transport of natural gases such as coal and oil release methane. The release of methane gas is also possible from livestock practices, agricultural practices and decomposition of organic waste present in the solid waste landfills.

Nitrous Oxide

Combustion of solid waste and fossil fuels releases nitrous oxide. Moreover, industrial and agricultural activities are the cause for release of this greenhouse gas.

Carbon Dioxide

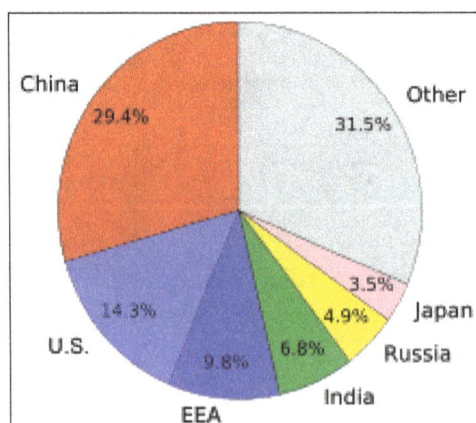

Major Carbon Dioxide Emitters.

The burning of fossil fuels such as coal, tar, and natural gas releases CO_2. Burning of waste materials, wood and trees release CO_2. Furthermore, the land use changes, deforestation, and soil degradation are the possible causes for adding up to the already existing CO_2 quantity.

Fluorinated Gases

Bunch of gases including hydrofluorocarbons, sulfur hexafluoride, and perfluorocarbons, etc are the fluorinated gases released during many commercial and industrial activities. Sometimes the gases are released due to household activities. However, these gases do not occur naturally. In certain cases, these chemicals are the replacement for ozone-depleting compounds like CFCs. However, they are equally harmful if not more.

Global Warming and Climate Change

The major climate change due to global warming is definitely a significant increase in global temperature. Moreover, the rise in temperature is the reason behind many deaths. It is also the cause of the increase in the incidence rate of some diseases such as dengue, malaria, etc.

Global warming is a threat to animals and plants. It is disturbing the entire ecological balance of the environment. Melting of polar ice caps and glaciers is increasing the rise in the water levels (sea and ocean) all around the globe. Furthermore, freshwater resources are reducing day by day.

According to scientific organizations like NASA, WHO and such, global warming is responsible for extreme weather incidences, ocean acidification, and many other similar impacts that are affecting the society and nature. However, it is possible to cease these drastic problems by adopting proper changes to our very own lifestyle.

Regulatory bodies and the governments across the globe should enforce proper policies to decrease the release of greenhouse gases in the atmosphere. Government and regulatory bodies like United Nations are actively trying to enforce policies to stop the effect.

Therefore, it is very important to participate in strong international agreements by countries across the globe and together work to save the planet and change the environment.

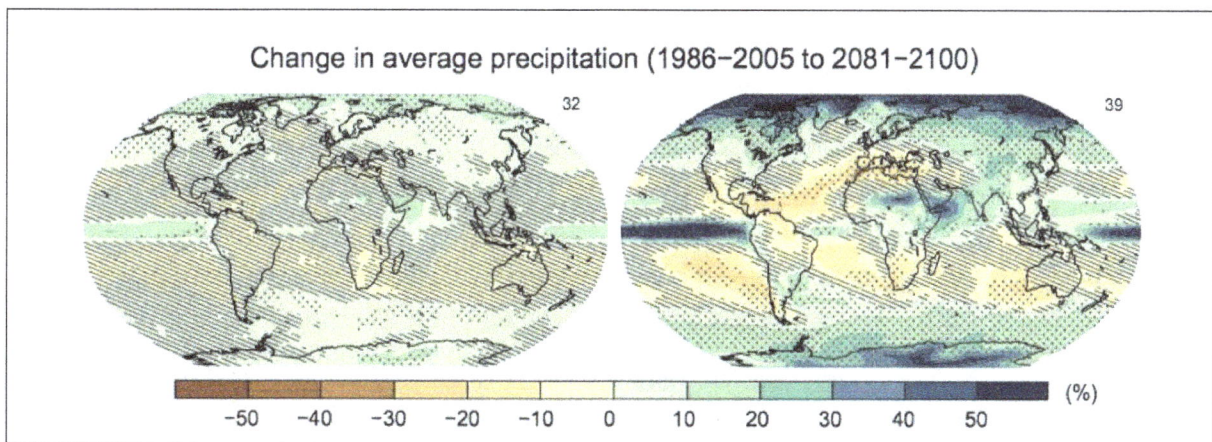

Change in Precipitation by End of 21st Century.

Photochemical Smog

Photochemical Smog Formation

The formation of photochemical smog involves three primary ingredients: nitrogen oxides, hydrocarbons and sunlight. The nitrogen oxides and hydrocarbons are by-products of fossil fuel-burning energy plants, and they can even come from natural processes, but the main source is the internal combustion engines in gasoline-powered automobiles.

Nitrous oxide and nitrogen dioxide dissociate in sunlight and combine with trace hydrocarbons to ultimately produce a large number of pollutants. The complex process proceeds in stages:

- Sunlight causes the photodissociation of nitrogen and oxygen to yield ozone and oxygen atoms.

- Oxygen atoms react with water to form hydroxyl radicals (OH).

- Hydroxyl radicals oxidize hydrocarbons to form hydrocarbon radicals.

- Hydrocarbons oxidize to form a class of chemicals known as aldehydes.

- Aldehydes oxidize to form aldehyde peroxides and aldehyde peroxyacids, which are the pollutants that create most of the health problems.

Chemicals in Photochemical Smog

Many major cities maintain a smog index, and one of the main chemical pollutants they monitor is ozone. It's produced as a byproduct of the dissociation of nitrogen compounds early in the process of smog formation, and although much of it gets used up in the formation of other pollutants, a significant amount doesn't. Ozone is corrosive. It causes respiratory ailments, and it damages plants, trees and even paint.

Besides ozone, photochemical smog contains a number of other pollutants, including:

- Peroxyacetyl nitrate (PAN): This pollutant causes eye and respiratory irritation and is primarily responsible for eye watering during periods of heavy air pollution.

- Nitrous acid (HNO_2): Mildly toxic, this compound also causes respiratory discomfort.

- Nitric acid (HNO_3): A stronger acid and one of the components of acid rain, nitric acid can burn skin and eyes in high concentrations. Nitric acid can also be formed during severe thunderstorms.

- Smog is a byproduct of modern industrialization: Due to industry and the number of motor vehicles, this is more of a problem in large cities that have a warm, sunny and dry climate.

- Oxidation: Photochemical smog is also referred to as oxidizing smog. Oxidation reactions have been defined several ways. In terms of oxygen transfer, oxidation is a gain of oxygen. Oxidation can also be defined as a loss of hydrogen. The most important use of oxidation is described in terms of electron transfer. Oxidation can be described as an increase in oxidation number or loss of electrons. Oxidation numbers represents a distribution of charge. In other words, oxidation numbers represent the charge of the atom if the compound was composed of ions. (Oxidation Numbers – Rules and Practice Problems).

- Reduction: Reduction can involve the gain of hydrogen or loss of oxygen. Reduction can refer to the gain of electrons, which results in a decrease in oxidation number.

Formation of Photochemical Smog

Step 1: People begin driving in the morning, nitrogen is burned or oxidized.

$$N_2 + O_2 \rightarrow 2NO$$

- Oxidation number of N_2 is 0. The nitrogen in NO has acquired an oxidation number of +2.

Step 2: After a few hours, NO combines with O_2, in another oxidation reaction.

$$2NO + O_2 \rightarrow 2NO_2$$

- The nitrogen in NO has an oxidation number of +2. The nitrogen in NO_2 has an oxidation number of +4.

Step 3: Nitrogen dioxide absorbs light energy, resulting in a reduction reaction.

$$NO_2 \rightarrow NO + O$$

- The nitrogen in NO_2 has an oxidation number of +4 and the nitrogen in NO is +2.

Step 4: In sunlight, atomic oxygen combines with oxygen gas to form ozone.

$$O + O_2 \rightarrow O_3$$

Step 5: Reaction is temperature and sunlight dependent.

$$O_3 + NO \rightleftharpoons NO_2 + O_2$$

Alternative Reactions

NO and NO_2 can also react with the hydrocarbons instead of ozone to form other volatile compounds known as PAN (peroxyacetyl nitrate). The accumulation of ozone and volatile organic compounds along with the energy from the sun forms the brown, photochemical smog seen on hot, sunny days.

Acid Rain

Acid rain is a broad term used to describe several ways that acids fall out of the atmosphere. A more precise term is acid deposition, which has two parts: wet and dry. Wet deposition refers to acidic rain, fog, and snow. As this acidic water flows over and through the ground, it affects a variety of plants and animals. Dry deposition refers to acidic gases and particles. About half of the acidity in the atmosphere falls back to earth through dry deposition. The wind blows these acidic particles and gases towards buildings, cars, homes and trees. Dry deposited gases and particles can also be washed from trees and other surfaces by rainstorms. When that happens, the runoff water adds those acids to the acid rain, making the combination more acidic than the falling rain alone. Precipitation that has a pH value of less than seven may contain acidic rain. This is due to the presence of acidic oxide emissions in the atmosphere from industries and vehicles. However, a rainfall that has a pH value of less than 5.6 is considered as acid rain. It is formed when sulphur dioxides and nitrogen oxides, as gases or fine reacts with rain water. Particles in the atmosphere combine with water vapour and precipitate as sulphuric acid or nitric acid in rain, snow, or fog. Therefore, the main objective of this paper was to assess the effect of acid rain on environment and to suggest the methods of preventing acid rain. Moreover, to review what have done on acid rain before and to forecast what will have done in the future. This is the first phase of the research. It will continued more on experimental result in the second phase of the paper.

Measuring Acid Rain

Acid rain is measured using pH meter from 1 to 14 value scales with a pH of 7.0 being neutral, 0 to 7 being acidic, and 7 to 14 basic. When the PH value lowers, the acidity nature of rain increases. Pure water has a pH value of 7. However, normal rain is slightly acidic because different acidic oxide emissions react with rain that lowers the pH value about 5.6. According to 2000 report, the most acidic rain falling in the US has a pH of about 4.3. This acid rain's pH and the chemicals that cause acid rain are monitored by two networks that are supported by EPA. The National Atmospheric Deposition Program measures wet deposition, and its Web site features maps of rainfall pH (follow the link to the isopleths maps) and other important precipitation chemistry measurements. The Clean Air Status and Trends Network (CASTNET) measures dry deposition. Its web site features information about the data it collects, the measuring sites, and the kinds of equipment it uses.

Areas Affected by Acid Rain due to Power Plant

Canada and USA: Acid rain is a problem in Eastern Canada and the Northeastern USA. Large smelters in western Ontario and steel processing plants in Indiana, Ohio historically used coal as a source of fuel. The sulfur dioxide produced was carried eastward by the jet stream. Acid rain from power plants in the Midwest United States has also harmed the forests of upstate New York and New England. In many areas water and soil systems lack natural alkalinity such as lime base cannot neutralize acid.

Sulfur dioxide is emitted from industrial processes and the burning of fossil fuels. In particular, ore smelting, coal-fired power generators, and the processing of natural gas result in the greatest emissions of sulfur dioxide. In 2000, Canada emitted 2.4 million tons of sulfur dioxide. Moreover, the primary causes of oxides of nitrogen are a vehicle, which accounts about 60% of all nitrogen oxide emissions. However, emissions also come from furnaces, boilers and engines. In 2000, Canada emitted 2.5 million tones of nitrogen oxide. Therefore, these emissions are the main causes of acid rain all over the world.

Europe and Asia: Industrial acid rain is a substantial problem in China, Eastern Europe and Russia and areas down-wind from them. The effects of acid rain can spread over a large area, far from the source of the pollution. Research carried out in North America in 1982, revealed that sulphur pollution killed 51,000 people and about 200,000 people become ill due to this emissions.

Over the past decades, Norway has suffered a great damage due to the effect of acid rain. While Norway's sulphur dioxide emissions have decreased significantly since the 1970s and 1980s, and nitrogen oxide emissions have decreased slightly, the damages from acid rain appear to be worsening in southern Norway. This is because it takes years for the ecosystems and the environment to recover from the effects of acidification. According to the State of the Environment in Norway, 18 salmon stocks have been lost and 12 are endangered, and have been wiped out of all of the large salmon rivers in southern Norway.

Hydrodesulphurization (HDS)

Hydro treating is a catalytic chemical process widely used to remove sulfur compounds from refined petroleum products such as gasoline or petrol, jet fuel, diesel fuel, and fuel oils. One purpose for removing the sulfur is to reduce the sulfur dioxide emissions resulting from using those fuels in automotive vehicles, aircraft, railroad locomotives, ships, or oil burning power plants, residential and industrial furnaces, and other forms of fuel combustion.

Another important reason for removing sulfur from the intermediate product naphtha streams within a petroleum refinery is that sulfur, even in extremely low concentrations, poisons the noble metal catalysts platinum and rhenium in the catalytic reforming units that are subsequently used to upgrade the of the naphtha streams.

Effects of Acid Rain on Environment

Harmful to aquatic life: This is due to increasing the acidity character in water bodies that Stops eggs of certain organisms (e.g. fish) to stop hatching, Changes population ratios and affects their ecosystem.

Harmful to vegetation: Vegetables are destructed due to increased acidity in soil, Leeches nutrients from soil, and slowing plant growth, poisoning plants, creates brown spots in leaves of trees, impeding photosynthesis, allows organisms to infect through broken leaves.

Affects human health: Causes respiratory problems, asthma, dry coughs, headaches and throat irritations; Leeching of toxins from the soil by acid rain can be absorbed by plants and animals. When consumed these toxins it affect human's life severely,which cause brain damage, kidney problems and Alzheimer's disease have been linked to people who eat meat of "toxic" animals/ plants by these pollutant.

Effect on transport: Currently, both the railway industry and the aeroplane industry have to spend a lot of money to repair the corrosive damage done by acid rain. Furthermore, bridges have collapsed in the past due to acid rain corrosion. Acid rain dissolves the stonework and mortar of buildings (especially those made out of sandstone or limestone). It reacts with the minerals in the stone to form a powdery substance that can be washed away by rain.

Preventing our Environment from Acidic Rain

There are several ways to reduce acid deposition and precipitation. These are:

Clean up smokestacks and exhaust pipes: Almost all of the electricity that powers modern life comes from burning fossil fuels like coal, natural gas, and oil. However, exhaust emission of these fuels are the main causes of acid deposition that released into the atmosphere. Coal fuel accounts for most US SO_2 and a large portion of NO_x emissions. Sulfur is present in coal as an impurity, and it reacts with air when the coal is burned to form SO_2. In contrast, NO_x is formed when any fossil fuel is burned. There are several options for reducing SO_2 emissions, including using coal containing less sulfur, washing the coal, and using devices called scrubbers to chemically remove the SO_2 from the gases leaving the smokestack and recycling to use as a raw material. Power plants can also switch fuels; for example burning natural gas creates much less SO_2 than burning coal. Certain approaches will also have additional benefits of reducing other pollutants such as mercury and carbon dioxide. Understanding these "co-benefits" has become important in seeking cost-effective air pollution reduction strategies. Finally, power plants can use technologies that don't burn fossil fuels. Each of these options has its own costs and benefits, however; there is no single universal solution. Similar to scrubbers on power plants, catalytic converters reduce NO_x emissions from cars. These devices have been required for over twenty years in the US, and it is important to keep them working properly and tailpipe restrictions have been tightened recently. EPA has also made, and continues to make, changes to gasoline that allows it to burn cleaner dioxide of sulfur (SO_2) and NO_x.

Use alternative energy sources: There are other sources of electricity besides fossil fuels such as nuclear power, hydropower, wind energy, geothermal energy, and solar energy. Of these, nuclear and hydropower are used most widely; wind, solar, and geothermal energy have not yet been harnessed on a large scale. There are also alternative energies available to power automobiles, including natural gas powered vehicles, battery-powered cars, fuel cells, biofuels and biodiesel and combinations of alternative and gasoline powered vehicles. All sources of energy have environmental costs as well as benefits. Some types of energy are more expensive to produce than others. Nuclear power, hydropower, and coal are the cheapest forms today, but changes in technologies and environmental regulations may shift that in the future. All of these factors must be weighed when deciding which energy source to use today and which to invest for tomorrow.

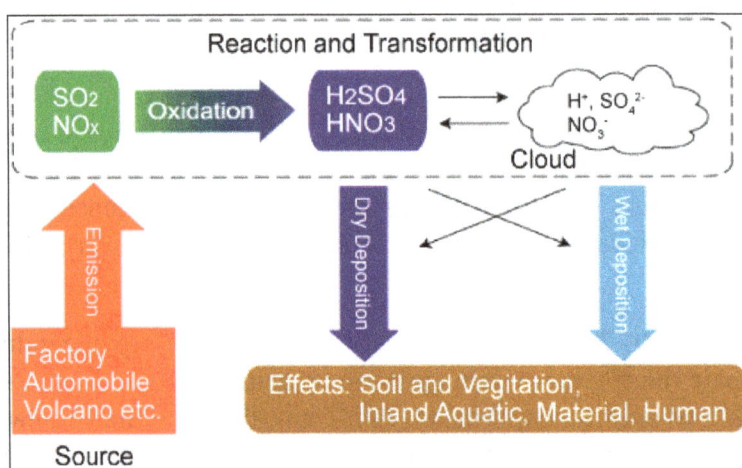

Flow chart that shows the formation of acid rain and its interaction with environment.

Liming: Powdered limestone added to water and soil to neutralize acid. It is commonly used in Norway and Sweden. However, it is more expensive and short-term remedy. Acid deposition penetrates deeply into the fabric of an ecosystem, changing the chemistry of the soil as well as the chemistry of the streams and narrowing, sometimes to nothing, the space where certain plants and animals can survive. Because there are so many changes, it takes many years for ecosystems to recover from acid deposition, even after emissions are reduced and the rain becomes normal again. For example, while the visibility might improve within days, and small or episodic chemical changes in streams improve within months, chronically acidified lakes, streams, forests, and soils can take years to decades or even centuries (in the case of soils) to heal. However, there are some things that people do to bring back lakes and streams more quickly. Limestone or lime (a naturally-occurring basic compound) can be added to acidic lakes to "cancel out" the acidity. This process, called liming. Liming tends to be expensive, has to be done repeatedly to keep the water from returning to its acidic condition, and is considered a short-term remedy in only specific areas rather than an effort to reduce or prevent pollution. Furthermore, it does not solve the broader problems of changes in soil chemistry and forest health in the watershed, and does nothing to address visibility reductions, materials damage, and risk to human health. However, liming does often permit fish to remain in a lake, so it allows the native population to survive in place until emissions reductions reduce the amount of acid deposition in the area.

Generally, rainfall that has a pH value less than 5.6 is considered as acid rain. It is formed when sulphur dioxides and nitrogen oxides reacted with water during rain and as gases or fine. Acids rain is described in terms of wet and dry depositions. The wet deposition refers to acidic rain, frog and snow whereas dry deposition refers to acidic gases and particles. This acid rain affects a variety of plants and animals (Harmful to aquatic life, Harmful to vegetation, affects human health and Transport) in our environment. We reduce it by Clean up smokestacks and exhaust pipes as wells as using alternative energy sources for vehicles and electricity generation for different purpose in order to live in a safe and suitable environment without fare of global warming.

References

- Ozone-depletion: britannica.com, Retrieved 16 July, 2019

- What-are-ozone-depleting-substances, risks-ozone-depleting-substances: mfe.govt.nz, Retrieved 23 May, 2019

- Greenhouse-effect-and-global-warming, environmental-chemistry: toppr.com, Retrieved 18 April, 2019

- Photochemical-smog-formed-6505511: sciencing.com, Retrieved 05 June, 2019

- Photochemical-Smog, Ancillary-Materials-Exemplars-and-Case-Studies-Exemplars-Environmental-and-Green-chemistry: chem.libretexts.org, Retrieved 25 March, 2019

- Mechanisms-and-effects-of-acid-rain-on-environment-2157-7617.1000204: omicsonline.org, Retrieved 18 May, 2019

Environmental Monitoring

Environmental Monitoring is the assessment, analysis, monitoring and protection of the quality of the environmental conditions by implementation and development of policies. It includes monitoring of air quality, water quality, soil quality, etc. All the aspects of environmental monitoring have been carefully analyzed in this chapter.

Environmental monitoring involves the assessment of the quality of the environment in order to control the risk of pollution.

It is used as the basis of the production of environmental impact assessments. This information is then used to study environmental trends and to quantify the current state of the environment.

Assessments through environmental monitoring are subject to statistical analysis. Environmental monitoring utilizes specialized equipment and specific monitoring methods to get the most accurate recording of pollutant levels.

Environmental monitoring takes the form of air, soil, and water monitoring. Air monitors are used to identify the air quality and the levels of pollution. Soil monitoring requires the collection of soil samples to be analyzed and tested in a laboratory setting. This may be by way of grab sampling or composite sampling. Water monitoring may take the form of chemical, biological, radiological, or microbiological testing.

The sampling method used depends on the type of environment, the sampling material, and the future use of the data collected.

Environmental monitoring is critical to the protection of human health and the environment. As the human population continues to increase, as industrial development and energy use continues to expand, and despite advances in pollution control, the continued production of pollution remains inevitable. Thus the need for environmental monitoring is still as great as ever. Continued advances in the development, application, and automation of monitoring devices are needed to enhance the accuracy and cost-effectiveness of monitoring programs. Equally as important is the need to produce more scientists and engineers that have the knowledge and training required to successfully develop and operate monitoring devices and manage monitoring programs.

Environmental monitoring data is often used in risk assessment to create environmental regulations. However, the task of interpreting data is critical in making decisions about potential health risks and corrective actions. Risk assessment is a process that has been formalized to estimate the risk of adverse health effects caused by exposure to harmful chemicals and microorganisms. The goal is to express risk caused by exposure to a contaminant in terms of probability of illness or mortality. In this format such information can be better utilized by decision makers to determine the magnitude of the problem and weigh the costs and benefits of prevention or corrective action.

Types of Sensors for Environmental Monitoring

There are numerous types of environmental sensors depending on the specifications needed for monitoring. Sensitivity, size and speed are all factors to consider when choosing an environmental sensor. Sensors can be divided into categories depending on the type of sensor being used, and the environmental factors that an analyst is looking at.

The main types of sensors include:

- Trace metal sensors: Laser-induced breakdown spectroscopy (LIBS), nanoelectrode array and miniature chemical flow-probe sensor.

- Radioisotopic sensors: RadFET (Radiation field-effect transistor), cadmium zinc telluride (CZT) detectors, low-energy pin diodes beta spectrometer, thermoluminescent dosimeter (TLD), isotope identification gamma detector, neutron generator for nuclear material detection, non-sandia radiation detectors.

- Volatile organic compound sensors: Evanescent fiber-optic chemical sensor, grating light reflection spectroscopy (GLRS), miniature chemical flow probe sensor, chemical sensor arrays, MicroChemLab (gas phase), gold nanoparticle chemiresistors, electrical impedance of tethered lipid bilayers on planar electrodes, MicroHound, hyperspectral imaging, chemiresistor array.

- Biological sensors: Fatty acid methyl esters (FAME) analyser, insulator-based dielectrophoresis, biological sensor arrays, µProLab, MicroChemLab (Liquid).

Sensors can also be split into types depending on environmental factors. These types of sensors include:

- Moisture sensors: Moisture sensors are essential for the measurement of volumetric water content in the soil towards the chimerical environment by several orders of magnitude.

- Pressure sensors/transducers: Pressure sensors operate on the basis of pressure applied. They are used in fields such as aviation, manufacturing biomedical measurements, auto mobile and hydraulic measurements.

- Tilt sensors: The tilt sensor relates the two different axes indicating a reference plane which may be in two different axes. This plays an important role in measuring tilt angles with reference to the Earth's ground plane. Tilt sensors are common in industry and in game controllers.

- Rain sensors: Rain sensors are a type of switching device which gets activated in the presence of rainfall. Rain sensors are used in areas such as as water conservation devices in irrigation systems, and in automobiles with windscreen wipers.

- Vibration sensors: Vibration sensors are important for displacement and acceleration with respect to the enduring impact on the environment. Displacement, linear velocity and acceleration are different factors which vibration sensors are measured according to.

Air Quality Monitoring

Air Quality Monitoring is the process of examining the quality of air in the atmosphere.

The starting point of air quality monitoring is to first study if an area has an air pollution problem. Monitoring helps in assessing the level of pollution in relation to the ambient air quality standards. Standards are a regulatory measure to set the target for pollution reduction and achieve clean air. Robust monitoring helps to guard against extreme events by alerting people and initiate action. We regulate a total of 12 pollutants, including SO_2, NO_2, PM10, PM2.5 (particulate matter of up to 10 micron and up to 2.5 micron size), ozone, lead, arsenic, nickel, CO, NH_3, benzene, and BaP (particulate phase). Across cities, only SO_2, NO_2 and RSPM / PM10 are monitored regularly. Other pollutants, such as PM2.5, O_3, CO, BTX, heavy metals are monitored in select cities as capacity is still being built.

Air Quality

"Air quality" refers to the condition of the air within our surrounding. Good air quality pertains to the degree which the air is clean, clear and free from pollutants such as smoke, dust and smog among other gaseous impurities in the air. Air quality is determined by assessing a variety of pollution indicators. Good air quality is a requirement for preserving the exquisite balance of life on earth for humans, plants, animals and natural resources. As such, human health, plants, animals and natural resources are threatened when pollution in the air reach high concentrations.

Poor air quality can affect or harm human health and/or the environment. Air quality can be degraded by natural or man-made sources. Natural sources include volcanic eruption, windstorm dust. Man-made source include pollution from moving vehicles, toxic gases from industries, coal powered plants, burning wood or other material in open air, landfills. Both these sources can seriously affect the overall air quality and can lead to severe health problems for humans.

Cause of Air Quality Degradation

Air quality is degraded by the day to day emissions from different air pollution sources. These include: stationary sources such as power plants, manufacturing facilities, smelters, motor machines, and burning wood and coal; mobile sources such as trucks, buses, planes and cars;

and natural sources such as volcanic eruptions, and windstorm dust. The pollutants from these sources are released into the air which can affect the health of humans, animals and our environment. The quality of air depends on 3 factors: (1) the amount of pollutants, (2) the rate at which they are released in the atmosphere and (3) how long they are trapped in an area.

The air pollution sources emit a wide range of pollutants classified under different categories that include Sulfur Dioxide (SO_2), particulate matter, Hydrocarbons (HC) and volatile organic compounds (VOC), Lead, Carbon Dioxide (CO_2) and Carbon Monoxide (CO), Nitrogen Oxides (NO_x), and Smog. In most regions, noticeable air quality degraders include smoke, dust, CO, NO_x, SO_2, HC, and smog from factories, power plants, automobiles, and smelters especially around urban and industrial areas.

Many of these air pollutants occur as gases, but some are very tiny solid particles, such as dust, smoke or soot. Air pollution from various sources can affect indoor air as well. Removing the pollutants from indoor air can improve the quality of your indoor air. Indoor air pollutants include cigarette smoke, mould, dust mites, pet dander, formaldehyde, volatile organic compounds (VOCs) and radon gas.

If air pollutants are in an area with good airflow, they will mix with the air and quickly disperse. Air pollutants tend to remain in the air when when there are certain conditions like light winds or mountains that restrict the transport of these pollutants away from an area. When this happens, pollution concentrations can increase rapidly.

Improving Air Quality

With all those toxic pollutants in the atmosphere, it isn't easy to get fresh as it used to be. Due to increase in number of vehicles and industries on a large scale, the air quality has gone from bad to worse in last few decades. Every year, millions of people across the globe die due to inhaling of toxic gases present in the atmosphere.

There are several measures which we can take to make air quality better not only for us but even for coming generations. If not done in proper and timely manner, the worsening air quality can cause severe harm to entire planetary ecosystem.

Dealing with Traffic and Vehicle Emissions

In a bid to improve air quality, dealing with automobile emission is the first choice for action. Beginning by reducing the numbers of automobile traffic and use of cleaner vehicles is the best course of action towards cutting down on emissions locally. Promoting public transport and proper urban planning can as well aid in reducing traffic numbers. Nowadays, journeys have become shorter and affordable due to introduction of fast electric trains and other forms of cheaper public transport.

By promoting public transport, it is easy to get people out of their vehicles which suggest that it's an obvious answer to reducing the number of cars on the road. Air pollution reduction can also be accomplished through speed reduction, promoting use of electric/hybrid cars, rerouting traffic, and introducing access restrictions. Through this, CO_2, lead, smoke, and particulate matter from road traffic emissions can considerably reduce thereby improving air quality. Besides use of cleaner vehicles and promotion of public transport, use of green fuels is a good alternative for improving air quality.

Dealing with Industrial Emissions

Factories and industrial manufacturing companies are another main concern when it comes to air pollution. Industries are the biggest air pollution sources as they emit high scores of pollutants into the air on daily basis. Due to the industrial emissions, the globe faces a great challenge of acid rain, ozone layer depletion, and global warming. Primary regulation policies by environmental agencies are required to regulate industrial emissions with an aim of large scale structural emission reduction from industrial facilities.

Control devices termed as "pollution eating nano-particles" can as well offer effective ways to reduce industrial air pollution by absorbing or destroying toxic emissions and contaminants from industries. It is realized through installation of particulate control devices in exhaust streams. Good examples include Titanium dioxide and ultraviolet light. Installed in an industrial facility, titanium dioxide reduces pollution by absorbing toxic emissions while ultraviolet light breaks up NO_x gases and VOCs. Other control devices include electrostatic precipitators, dust cyclones, particulate scrubbers, and bag-houses. Switching to use of renewable energy is the most effective mechanism for improving air quality standards in industrial manufacturing.

Establishment of Policies and Clean Air Action Plans

Air quality improvement equally requires a string of directives, support, and initiatives geared towards emission reduction not only locally but internationally as well. Local, national and international environmental protection agencies need to work hand in hand to ensure the set air quality levels are met. It's a commitment that can only be achieved through an effective environmental action plan, strategies and policies directed towards addressing the underlying air pollution challenges. For instance, the Clean Air for Europe (CAFE) and the Environmental Protection Agency (EPA) programs have set standards for restricting air pollution and it has since resulted to a new directive for clean and quality air. Established about a decade ago, the directives have managed to set in place emission goals and standards for air quality.

Objectives are set whereby all sorts of air pollutants have to be regulated and cut down to a standard level and also managed within the scope of the challenges experienced. In challenging situations, deadlines for complying must be instituted and upheld. As such, creation of action plan directives similar to CAFE's, EPA's or any other devoted environmental protection agency can offer a wonderful advancement in the direction of improving air quality.

To work towards the attainment of improving the air quality situation, each state and nation without air regulation policies can make it a requirement to develop an action plan for cleaning the air within their jurisdiction. In addition to development of an action plan, implementation is essential as it defines the efforts that will improve and attempt to regulate the air pollution levels and to improve the air quality of the region.

Also through various environmental protection and clean air support programs, states and nation can establish a monitoring criteria for pollutants. One similar program is the Ambient Air Monitoring that has been adopted by various nations and states. Through the program, statistics can be collected regarding air quality by measuring samples so as to estimate the attainment of the set air quality standards. In this sense, pollution trends can be observed throughout the jurisdiction region to deter or reduce air pollution crisis. Other air quality monitoring programs can encompass Air Pollution Monitoring and Enhanced Ozone Monitoring.

Air pollution reduction efforts can also involve land-use, urban, and transportation planning strategies. The utilization of these strategies can assess the root causes of air pollution and to install measures for dealing with the challenges experienced.

Accordingly, environmental clean air policies and programs can be established to address issues regarding quality of air as a factor directly impacting on human health and the environment. The world, particularly regions burdened with the air pollution challenge, can thus improve air quality by adopting or establishing clean air action plans and support programs.

Air Quality Index

Air quality index (AQI) is a numerical scale used for reporting day to day air quality with regard to human health and the environment. The daily results of the index are used to convey to the public an estimate of air pollution level. An increase in air quality index signifies increased air pollution and severe threats to human health. In most cases, AQI indicates how clear or polluted the air in our surrounding is, and the associated health risks it might present. The AQI centers on the health effects that may be experienced within a few days or hours after breathing polluted air.

AQI calculations focus on major air pollutants including: particulate matter, ground-level ozone, sulfur dioxide (SO_2), nitrogen dioxide (NO_2), and carbon monoxide (CO). Particulate matter and ozone pollutants pose the highest risks to human health and the environment. For each of these air pollutant categories, different countries have their own established air quality indices in relation to other nationally set air quality standards for public health protection.

Air Quality Determination

On an hourly basis, the concentration of each pollutant in the air is measured and converted into a number running from zero upwards by using a standard index or scale. Calculated number for every pollutant is termed as a sub-index. The highest sub-index for any given hour is recorded as the AQI for that hour. In simple terms, AQI is like a yardstick that ranges from zero to five hundred (0-500). The index is a relative scale, meaning, the lower the index the better the quality of air and the lesser the health concern, and vice versa. The concentration of each pollutant varies, therefore, AQI values are grouped into ranges assigned to a standardized public health warnings and color code.

For instance, an AQI value of 0 to 50 means good air quality with limited possibility of affecting public and environmental health. On the other hand, an AQI value of 300 to 500 represents hazardous air quality with greater potential to affect public and environment health. The commonly accepted value is 100 as it corresponds to the generally approved air quality standards as set to safeguard public health. AQI levels below 100 are highly satisfactory while values beyond 100 are regarded to harm human health. As the AQI values get higher, it even poses more serious health concerns.

During days that the AQI is recorded to be elevated, the institution of public health might: offer advice to sensitive groups such as the children, those with respiratory problems, and the elderly to keep away from outdoor activities; take action to improve air quality by introducing measures for reducing emissions; or recommend the use of air pollution masks, especially in severe cases of air pollution.

Formula for Calculating Air Quality Index

Air quality index is calculated by a linear function by determining the concentration of the pollutant. The equation below is used to compute AQI.

$$I = \frac{I_{high} - I_{low}}{C_{high} - C_{low}} (C - C_{low}) + I_{low}$$

Where:

- I = the (Air Quality) index,

- C = the pollutant concentration,

- C_{low} = the concentration breakpoint that is ≤ C,

- C_{high} = the concentration breakpoint that is ≥ C,

- I_{low} = the index breakpoint corresponding to C_{low},

- I_{high} = the index breakpoint corresponding to C_{high}.

Air Quality Index Categories

The AQI is divided in six categories and each category is meant to correspond to different health concern levels. Below is an explanation of the categories and their meanings.

- 0 – 50 indicates "Good" AQI. At this level the quality of air is deemed to be satisfactory, and air pollution poses little or no risk.

- 51 – 100 indicate "Moderate" AQI. This means acceptable Air quality. However, some pollutants might arouse modest health concern for a limited number of people. For instance, persons who are remarkably sensitive to ozone may experience respiratory symptoms.

- 101 – 150 indicate "Unhealthy for Sensitive Groups" AQI. This category may not be able to affect the general health of the public. However, children, older adults, and persons with lung disease are at a greater risk from ozone exposure. Older children, adults and people with lung and heart disease are at greater risk from exposure to particulate matter.

- 151 – 200 indicate "Unhealthy" AQI. In this category, every person might experience some adverse health effects. Sensitive group members may experience more serious effects. People with heart or lung disease, older adults and children should cut back or reschedule strenuous activities.

- 201 – 300 indicate "Very Unhealthy" AQI. This would issue a health alert to mean that everybody may experience very serious health implications. People with heart or lung disease, older adults and children should significantly cut back or reschedule strenuous activities.

- Greater than 300 indicate "Hazardous" AQI. Air quality at this level is life-threatening and would issue warnings of emergency conditions for the entire population.

Table: Air Quality Index (AQI) categories.

Air Quality Index (AQI) Values	Levels of Health Concern	Colors
When the AQI is in this range:	air quality conditions are:	as symbolized by this color:
0-50	Good	Green
51-100	Moderate	Yellow
101-150	Unhealthy for Sensitive Groups	Orange
151 to 200	Unhealthy	Red
201 to 300	Very Unhealthy	Purple
301 to 500	Hazardous	Maroon

The AQI is calculated for four major air pollutants regulated by the Clean Air Act: ground-level ozone, particle pollution, carbon monoxide, and sulfur dioxide. You need to take following simple steps to avoid exposure to unhealthy air:

Prolonged Exertion: Prolonged exertion is any outdoor activity that you do intermittently for several hours and may cause you to breather slightly faster than normal. When air is unhealthy outside, you can reduce intake of unhealthy air by reducing how much time you spend on this type of activity.

Heavy Exertion: Heavy exertion means intense outdoor activities that cause you to breathe hard. When air quality is bad outside, you can protect your health by reducing the amount of time you spend on this activity or by substituting it with less intense activity.

Computation

Computation of the AQI requires an air pollutant concentration over a specified averaging period, obtained from an air monitor or model. Taken together, concentration and time represent the dose of the air pollutant. Health effects corresponding to a given dose are established by epidemiological research. Air pollutants vary in potency, and the function used to convert from air pollutant concentration to AQI varies by pollutant. Its air quality index values are typically grouped into ranges. Each range is assigned a descriptor, a color code, and a standardized public health advisory.

An air quality measurement station.

The AQI can increase due to an increase of air emissions (for example, during rush hour traffic or when there is an upwind forest fire) or from a lack of dilution of air pollutants. Stagnant air, often caused by an anticyclone, temperature inversion, or low wind speeds lets air pollution remain in a local area, leading to high concentrations of pollutants, chemical reactions between air contaminants and hazy conditions.

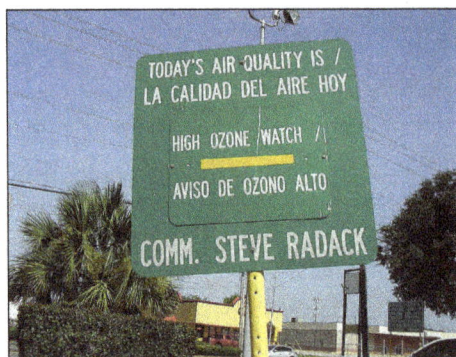

Signboard indicating an ozone watch.

On a day when the AQI is predicted to be elevated due to fine particle pollution, an agency or public health organization might:

- Advise sensitive groups, such as the elderly, children, and those with respiratory or cardiovascular problems, to avoid outdoor exertion.

- Declare an "action day" to encourage voluntary measures to reduce air emissions, such as using public transportation.

- Recommend the use of masks to keep fine particles from entering the lungs.

During a period of very poor air quality, such as an air pollution episode, when the AQI indicates that acute exposure may cause significant harm to the public health, agencies may invoke emergency plans that allow them to order major emitters (such as coal burning industries) to curtail emissions until the hazardous conditions abate.

Most air contaminants do not have an associated AQI. Many countries monitor ground-level ozone, particulates, sulfur dioxide, carbon monoxide and nitrogen dioxide, and calculate air quality indices for these pollutants.

The definition of the AQI in a particular nation reflects the discourse surrounding the development of national air quality standards in that nation. A website allowing government agencies anywhere in the world to submit their real-time air monitoring data for display using a common definition of the air quality index has recently become available.

Air Quality Sampling

There are 5 main methods of sampling air quality:

Passive Monitoring

- Diffusion tubes absorb a specific pollutant from the ambient air - no power supply is needed.

- Diffusion tubes normally monitor for 2-4 weeks at a time.

- The tubes must be sent to a laboratory for analysis to see how much pollution they have detected.

Active Sampling

- An analyser pulls the ambient air through a filter for a set period of time e.g. one filter per day.

- The filters are then collected and sent to a laboratory for analysis to see how much pollution they have detected.

Automatic Point Monitoring

- Ambient air is pulled through an analyser which recognises the chosen gas and will calculate its concentration.

- Automatic sites monitor pollutants 24hours a day.

- Data is sent from the site directly to your computer which means it can be seen instantly.

Photochemical and Optical Sensor Systems

- These are portable monitoring tools that can continuously monitor a range of pollutants. The sensors are of low sensitivity and mostly suitable for identifying hotspots at roadsides and near point sources.

- Data can be downloaded to your computer and analysed.

Remote Optical Monitoring

- This method of sampling detects pollution between a a light source and a detector which are placed separately at a site.

- Real time measurements can be taken with this type of sampling.

- Data can be sent from the analyser directly to your computer which means it can be seen instantly.

Water Quality Monitoring

Water quality monitoring aims to obtain qualitative information on the physical, chemical, and biological characteristics of a water body.

The objectives of a water quality monitoring program can be manifold. They range from plain monitoring of the environmental state of a water body to the detection of drinking water standard violations, from the analysis of temporary water quality trends to the observation of an industrial plant's injection of waste water. Last not least maintaining and monitoring a certain water quality level is made compulsory by many national and international regulations.

Deteriorating quality of water, for whichever reason, can pose a major hazard to human health, aquatic live and entire eco-systems. Timely detection is of crucial importance to minimize and mitigate the impact of such events, and/or to determine the source of pollution. Particularly in rivers and streams short polluting events, such as illegal disposal of toxic wastes or excessive waste water loads, are nearly impossible to detect by taking samples. Timely detection of such events can help taking preventive action as the wave of pollutants moves downstream.

The situation is aggravated by the various parameters that need to be monitored. While for many cases simple indicators such as temperature and conductivity will suffice, in other cases more sophisticated parameters, such as pH, dissolved oxygen (DO), turbidity or blue algae content will need to be monitored.

Water Quality Monitoring Data

Water quality monitoring data is incredibly useful however it's not always easy to gather. Specialists use a range of different techniques to put together results, including taking samples of chemical conditions, analysing sediments and using fish tissue extracts to find traces of metals, oils, pesticides, dissolved oxygen and nutrients. Physical conditions such as temperature, erosion and flow offer valuable insight while biological measurements regarding plant and animal life indicate the health of aquatic ecosystems.

At the end of the day, water quality monitoring is an essential part of keeping the planet healthy and sustainable. As we continue to build cities, clear land for farming and make other man-made changes to the natural environment, water quality monitoring becomes increasingly important. Land based activities can have a huge impact on water systems and it's critical that we realise how these affect waterbodies, both above and below ground.

Water Quality

Water quality is measured by several factors, such as the concentration of dissolved oxygen, bacteria levels, the amount of salt (or salinity), or the amount of material suspended in the water (turbidity). In some bodies of water, the concentration of microscopic algae and quantities of pesticides, herbicides, heavy metals, and other contaminants may also be measured to determine water quality.

Although scientific measurements are used to define water quality, it is not a simple thing to say "that water is good" or "that water is bad." So, the determination is typically made relative to the purpose of the water – is it for drinking or to wash a car with or for some other purpose?

Poor water quality can pose a health risk for people. Poor water quality can also pose a health risk for ecosystems.

The parameters for water quality are determined by the intended use. Work in the area of water quality tends to be focused on water that is treated for human consumption, industrial use, or in the environment.

Human Consumption

Contaminants that may be in untreated water include microorganisms such as viruses, protozoa and bacteria; inorganic contaminants such as salts and metals; organic chemical contaminants from industrial processes and petroleum use; pesticides and herbicides; and radioactive contaminants. Water quality depends on the local geology and ecosystem, as well as human uses such as sewage dispersion, industrial pollution, use of water bodies as a heat sink, and overuse (which may lower the level of the water).

The United States Environmental Protection Agency (EPA) limits the amounts of certain contaminants in tap water provided by US public water systems. The Safe Drinking Water Act authorizes EPA to issue two types of standards:

- Primary standards regulate substances that potentially affect human health;

- Secondary standards prescribe aesthetic qualities, those that affect taste, odor, or appearance.

The U.S. Food and Drug Administration (FDA) regulations establish limits for contaminants in bottled water that must provide the same protection for public health. Drinking water, including bottled water, may reasonably be expected to contain at least small amounts of some contaminants. The presence of these contaminants does not necessarily indicate that the water poses a health risk.

In urbanized areas around the world, water purification technology is used in municipal water systems to remove contaminants from the source water (surface water or groundwater) before it is distributed to homes, businesses, schools and other recipients. Water drawn directly from a stream, lake, or aquifer and that has no treatment will be of uncertain quality.

Industrial and Domestic Use

Dissolved minerals may affect suitability of water for a range of industrial and domestic purposes. The most familiar of these is probably the presence of ions of calcium (Ca^{2+}) and magnesium (Mg^{2+}) which interfere with the cleaning action of soap, and can form hard sulfate and soft carbonate deposits in water heaters or boilers. Hard water may be softened to remove these ions. The softening process often substitutes sodium cations. Hard water may be preferable to soft water for human consumption, since health problems have been associated with excess sodium and with calcium and magnesium deficiencies. Softening decreases nutrition and may increase cleaning effectiveness. Various industries' wastes and effluents can also pollute the water quality in receiving bodies of water.

Environmental Water Quality

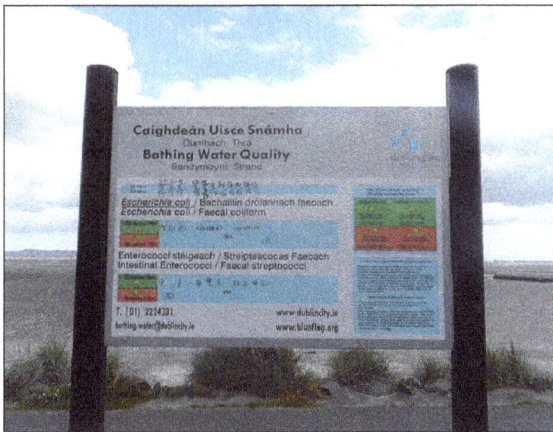

Sign describing water quality, giving levels of faecal coliform E. coli and Enterococcus faecalis.

Urban runoff discharging to coastal waters.

Environmental water quality, also called ambient water quality, relates to water bodies such as lakes, rivers, and oceans. Water quality standards for surface waters vary significantly due to different environmental conditions, ecosystems, and intended human uses. Toxic substances and high populations of certain microorganisms can present a health hazard for non-drinking purposes such as irrigation, swimming, fishing, rafting, boating, and industrial uses. These conditions may also affect wildlife, which use the water for drinking or as a habitat. Modern water quality laws generally specify protection of fisheries and recreational use and require, as a minimum, retention of current quality standards.

There is some desire among the public to return water bodies to pristine, or pre-industrial conditions. Most current environmental laws focus on the designation of particular uses of a water body. In some countries these designations allow for some water contamination as long as the particular type of contamination is not harmful to the designated uses. Given the landscape changes (e.g., land development, urbanization, clearcutting in forested areas) in the watersheds of many freshwater bodies, returning to pristine conditions would be a significant challenge. In these cases, environmental scientists focus on achieving goals for maintaining healthy ecosystems and may concentrate on the protection of populations of endangered species and protecting human health.

Satirical cartoon by William Heath, showing a woman
observing monsters in a drop of water.

Sampling and Measurement

The complexity of water quality as a subject is reflected in the many types of measurements of water quality indicators. The most accurate measurements of water quality are made on-site, because water exists in equilibrium with its surroundings. Measurements commonly made on-site and in direct contact with the water source in question include temperature, pH, dissolved oxygen, conductivity, oxygen reduction potential (ORP), turbidity, and Secchi disk depth.

Sample Collection

An automated sampling station installed along the East Branch Milwaukee River, New Fane, Wisconsin. The cover of the 24-bottle autosampler (center) is partially raised, showing the sample bottles inside. The autosampler was programmed to collect samples at time intervals, or proportionate to flow over a specified period. The data logger (white cabinet) recorded temperature, specific conductance, and dissolved oxygen levels.

More complex measurements are often made in a laboratory requiring a water sample to be collected, preserved, transported, and analyzed at another location. The process of water sampling introduces two significant problems:

- The first problem is the extent to which the sample may be representative of the water source of interest. Many water sources vary with time and with location. The measurement of interest may vary seasonally or from day to night or in response to some activity of man or natural populations of aquatic plants and animals. The measurement of interest may vary with distances from the water boundary with overlying atmosphere and underlying or confining soil. The sampler must determine if a single time and location meets the needs of the investigation, or if the water use of interest can be satisfactorily assessed by averaged values with time and location, or if critical maxima and minima require individual measurements over a range of times, locations or events. The sample collection procedure must assure correct weighting of individual sampling times and locations where averaging is appropriate. Where critical maximum or minimum values exist, statistical methods must be applied to observed variation to determine an adequate number of samples to assess probability of exceeding those critical values.

- The second problem occurs as the sample is removed from the water source and begins to establish chemical equilibrium with its new surroundings – the sample container. Sample containers must be made of materials with minimal reactivity with substances to be measured; and pre-cleaning of sample containers is important. The water sample may dissolve part of the sample container and any residue on that container, or chemicals dissolved in the water sample may sorb onto the sample container and remain there when the water is poured out for analysis. Similar physical and chemical interactions may take place with any pumps, piping, or intermediate devices used to transfer the water sample into the sample container. Water collected from depths below the surface will normally be held at the reduced pressure of the atmosphere; so gas dissolved in the water may escape into unfilled space at the top of the container. Atmospheric gas present in that air space may also dissolve into the water sample. Other chemical reaction equilibria may change if the water sample changes temperature. Finely divided solid particles formerly suspended by water turbulence may settle to the bottom of the sample container, or a solid phase may form from biological growth or chemical precipitation. Microorganisms within the water sample may biochemically alter concentrations of oxygen, carbon dioxide, and organic compounds. Changing carbon dioxide concentrations may alter pH and change solubility of chemicals of interest. These problems are of special concern during measurement of chemicals assumed to be significant at very low concentrations.

Filtering a manually collected water sample for analysis.

Sample preservation may partially resolve the second problem. A common procedure is keeping samples cold to slow the rate of chemical reactions and phase change, and analyzing the sample as soon as possible; but this merely minimizes the changes rather than preventing them. A useful procedure for determining influence of sample containers during delay between sample collection and analysis involves preparation for two artificial samples in advance of the sampling event. One sample container is filled with water known from previous analysis to contain no detectable amount of the chemical of interest. This sample, called a "blank", is opened for exposure to the atmosphere when the sample of interest is collected, then resealed and transported to the laboratory with the sample for analysis to determine if sample holding procedures introduced any measurable amount of the chemical of interest. The second artificial sample is collected with the sample of interest, but then "spiked" with a measured additional amount of the chemical of interest at the time of collection. The blank and spiked samples are carried with the sample of interest and analyzed by the same methods at the same times to determine any changes indicating gains or losses during the elapsed time between collection and analysis.

Testing in Response to Natural Disasters and other Emergencies

Inevitably after events such as earthquakes and tsunamis, there is an immediate response by the aid agencies as relief operations get underway to try and restore basic infrastructure and provide the basic fundamental items that are necessary for survival and subsequent recovery. Access to clean drinking water and adequate sanitation is a priority at times like this. The threat of disease increases hugely due to the large numbers of people living close together, often in squalid conditions, and without proper sanitation.

After a natural disaster, as far as water quality testing is concerned there are widespread views on the best course of action to take and a variety of methods can be employed. The key basic water quality parameters that need to be addressed in an emergency are bacteriological indicators of fecal contamination, free chlorine residual, pH, turbidity and possibly conductivity/total dissolved solids. There are a number of portable water test kits on the market widely used by aid and relief agencies for carrying out such testing.

After major natural disasters, a considerable length of time might pass before water quality returns to pre-disaster levels. For example, following the 2004 Indian Ocean tsunami the Colombo-based International Water Management Institute (IWMI) monitored the effects of saltwater and concluded that the wells recovered to pre-tsunami drinking water quality one and a half years after the event. IWMI developed protocols for cleaning wells contaminated by saltwater; these were subsequently officially endorsed by the World Health Organization as part of its series of Emergency Guidelines.

Chemical Analysis

A gas chromatograph- mass spectrometer measures pesticides and other organic pollutants.

The simplest methods of chemical analysis are those measuring chemical elements without respect to their form. Elemental analysis for oxygen, as an example, would indicate a concentration of 890 g/L (grams per litre) of water sample because oxygen (O) has 89% mass of the water molecule (H_2O). The method selected to measure dissolved oxygen should differentiate between diatomic oxygen and oxygen combined with other elements. The comparative simplicity of elemental analysis has produced a large amount of sample data and water quality criteria for elements sometimes identified as heavy metals. Water analysis for heavy metals must consider soil particles suspended in the water sample. These suspended soil particles may contain measurable amounts of metal. Although the particles are not dissolved in the water, they may be consumed by people drinking the water. Adding acid to a water sample to prevent loss of dissolved metals onto the sample container may dissolve more metals from suspended soil particles. Filtration of soil particles from the water sample before acid addition, however, may cause loss of dissolved metals onto the filter. The complexities of differentiating similar organic molecules are even more challenging.

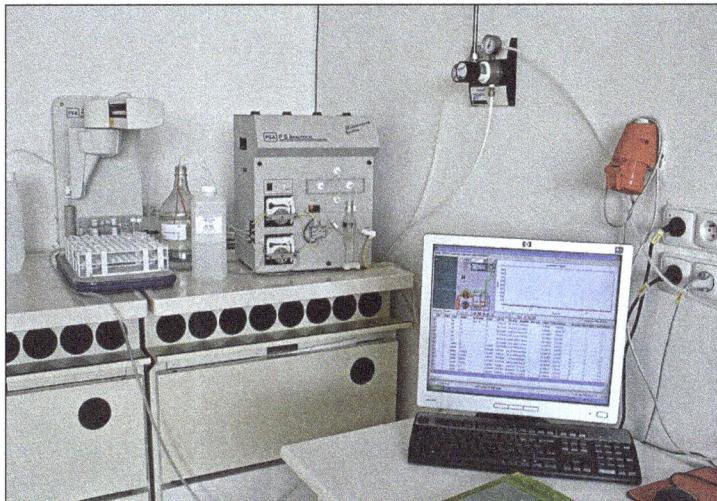

Atomic fluorescence spectroscopy is used to measure mercury and other heavy metals.

Making these complex measurements can be expensive. Because direct measurements of water quality can be expensive, ongoing monitoring programs are typically conducted by government agencies. However, there are local volunteer programs and resources available for some general assessment. Tools available to the general public include on-site test kits, commonly used for home fish tanks, and biological assessment procedures.

Real-time Monitoring

Although water quality is usually sampled and analyzed at laboratories, since the late 20th century there has been increasing public interest in the quality of drinking water provided by municipal systems. Many water utilities have developed systems to collect real-time data about source water quality. In the early 21st century, a variety of sensors and remote monitoring systems have been deployed for measuring water pH, turbidity, dissolved oxygen and other parameters. Some remote sensing systems have also been developed for monitoring ambient water quality in riverine, estuarine and coastal water bodies.

Drinking Water Indicators

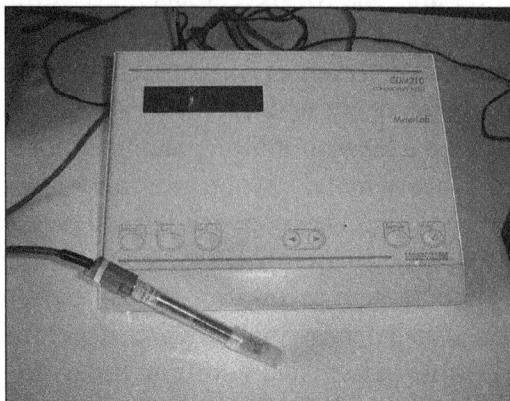

An electrical conductivity meter is used to measure total dissolved solids.

The following is a list of indicators often measured by situational category:

- Alkalinity;

- Color of water;

- pH;

- Taste and odor (geosmin, 2-Methylisoborneol (MIB), etc.);

- Dissolved metals and salts (sodium, chloride, potassium, calcium, manganese, magnesium);

- Microorganisms such as fecal coliform bacteria (*Escherichia coli*), Cryptosporidium, and Giardia lamblia; *see* Bacteriological water analysis;

- Dissolved metals and metalloids (lead, mercury, arsenic, etc.);

- Dissolved organics: colored dissolved organic matter (CDOM), dissolved organic carbon (DOC);

- Radon;

- Heavy metals;

- Pharmaceuticals;

- Hormone analogs.

Environmental Indicators

Physical Indicators

- Water temperature;

- Specific conductance or electrical conductance (EC) or conductivity;

- Total suspended solids (TSS);

- Transparency or turbidity;

- Total dissolved solids (TDS);

- Odour of water;

- Color of water;

- Taste of water.

Chemical Indicators

- pH;

- Biochemical oxygen demand (BOD);

- Chemical oxygen demand (COD);

- Dissolved oxygen (DO);

- Total hardness (TH);

- Heavy metals;

- Nitrate;

- Orthophosphates;

- Pesticides;

- Surfactants.

Biological Indicators

- *Ephemeroptera*;

- *Plecoptera*;

- *Mollusca*;

- *Trichoptera*;

- *Escherichia coli* (E. coli);

- *Coliform* bacteria;

- *Pimephales promelas* (fathead minnow);

- *Americamysis bahia* (Mysid shrimp);

- *Sea urchin.*

Biological monitoring metrics have been developed in many places, and one widely used measure is the presence and abundance of members of the insect orders Ephemeroptera, Plecoptera and Trichoptera (common names are, respectively, mayfly, stonefly and caddisfly). EPT indexes will naturally vary from region to region, but generally, within a region, the greater the number of taxa from these orders, the better the water quality. Organisations in the United States, such as EPA. offer guidance on developing a monitoring program and identifying members of these and other aquatic insect orders. Many US wastewater dischargers (e.g., factories, power plants, refineries, mines, municipal sewage treatment plants) are required to conduct periodic whole effluent toxicity (WET) tests.

Individuals interested in monitoring water quality who cannot afford or manage lab scale analysis can also use biological indicators to get a general reading of water quality. One example is the IO-WATER volunteer water monitoring program of Iowa, which includes a benthic macroinvertebrate indicator key.

Bivalve molluscs are largely used as bioindicators to monitor the health of aquatic environments in both fresh water and the marine environments. Their population status or structure, physiology, behaviour or the level of contamination with elements or compounds can indicate the state of con-tamination status of the ecosystem. They are particularly useful since they are sessile so that they are representative of the environment where they are sampled or placed. A typical project is the U.S. Mussel Watch Programme, but today they are used worldwide.

The Southern African Scoring System (SASS) method is a biological water quality monitoring system based on the presence of benthic macroinvertebrates. The SASS aquatic biomonitor-ing tool has been refined over the past 30 years and is now on the fifth version (SASS5) which has been specifically modified in accordance with international standards, namely the ISO/IEC 17025 protocol. The SASS5 method is used by the South African Department of Water Affairs as a standard method for River Health Assessment, which feeds the national River Health Programme and the national Rivers Database.

Water Quality Index

A water quality index provides a single number (like a grade) that expresses overall water quality at a certain location and time based on several water quality parameters. The objective of an index is to turn complex water quality data into information that is understandable and useable by the public. This type of index is similar to the index developed for air quality that shows if it's a red or blue air quality day. The use of an index to "grade" water quality is a controversial issue among wa-ter quality scientists. A single number cannot tell the whole story of water quality; there are many other water quality parameters that are not included in the index. The index presented here is not specifically aimed at human health or aquatic life regulations. However, a water index based on some very important parameters can provide a simple indicator of water quality. It gives the public a general idea the possible problems with the water in the region.

Soil Quality Monitoring

Soil quality is best defined in relation to the functions that soils perform in natural and agro-ecosystems. The quality of soil resources has historically been closely related to soil productivity. Indeed, in many cases the terms soil quality and soil productivity have been nearly synonymous.

More recently, however, there is growing recognition that the functions soils carry out in natural and agroecosystems go well beyond promoting the growth of plants. The need to broaden the concept of soil quality beyond traditional concerns for soil productivity have been highlighted at a series of recent conferences and symposia.

Johnson and colleagues, in a paper presented at a Symposium on Soil Quality Standards hosted by the Soil Science Society of America in October 1990 suggested that soil quality should be defined in terms of the function soils play in the environment and defined soil function as "the potential utility of soils in landscapes resulting from the natural combination of soil chemical, physical, and biological attributes". They recommended that policies to protect soil resources should protect the soil's capacity to serve several functions simultaneously including the production of food, fiber and fuel; nutrient and carbon storage; water filtration, purification, and storage; waste storage and degradation; and the maintenance of ecosystem stability and resiliency.

Larson and Pierce defined soil quality as "the capacity of a soil to function, both within its ecosystem boundaries (e.g., soil map unit boundaries) and with the environment external to that ecosystem (particularly relative to air and water quality)". They proposed "fitness for use" as a simple operational definition of soil quality and stressed the need to explicitly address the function of soils as a medium for plant growth, in partitioning and regulating the flow of water in the environment, and as an environmental buffer. Parr and colleagues, defined soil quality as "the capability of a soil to produce safe and nutritious crops in a sustained manner over the long-term, and to enhance human and animal health, without impairing the natural resource base or harming the environment". Parr and colleagues stressed the need to expand the notion of soil quality beyond soil productivity to include the role of the soil as an environmental filter affecting both air and water quality. They suggested that soil quality has important effects on the nutritional quality of the food produced in those soils but noted that these linkages are not well understood and research is needed to clarify the relationship between soil quality and the nutritional quality of food.

There is a growing recognition of the importance of the functions soils perform in the environment. The importance of those functions requires that scientists, policymakers, and producers adopt a broader definition of soil quality. Soil quality is best defined as the capacity of a soil to promote the growth of plants; protect watersheds by regulating the infiltration and partitioning of precipitation; and prevent water and air pollution by buffering potential pollutants such as agricultural chemicals, organic wastes, and industrial chemicals. The quality of a soil is determined by a combination of physical, chemical, and biological properties such as texture, water-holding capacity, porosity, organic matter content, and depth. Since these attributes differ among soils, soils differ in their quality. Some soils, because of their texture or depth, for example, are inherently more productive because they can store and make available larger amounts of water and nutrients to plants. Similarly, some soils, because of their organic matter content, are able to immobilize or degrade larger amounts of potential pollutants.

Soil management can either improve or degrade soil quality. Erosion, compaction, salinization, sodification, acidification, and pollution with toxic chemicals can and do degrade soil quality. Increasing soil protection by crop residues and plants; adding organic matter to the soil through crop rotations, manures, or crop residues; and careful management of fertilizers, pesticides, tillage equipment, and other elements of the farming system can improve soil quality.

Importance of Soil Quality

Soils have important direct and indirect impacts on agricultural productivity, water quality, and the global climate. Soils make it possible for plants to grow by mediating the biological, chemical, and physical processes that supply plants with nutrients, water, and other elements. Microorganisms in soils transform nutrients into forms that can be used by growing plants. Soils are the storehouses for water and nutrients. Plants draw on these stores as needed to produce roots, stems, leaves, and, eventually, food and fiber for human consumption. Soils—and the biological, chemical, and physical processes they make possible—are a fundamental resource on which the productivities of agricultural and natural ecosystems depend.

The soil, which interacts with landscape features and plant cover, is a key element in regulating and partitioning water flow through the environment. Rainfall in terrestrial ecosystems falls on the soil surface where it either infiltrates the soil or moves across the soil surface into streams or lakes. The condition of the soil surface determines whether rainfall infiltrates or runs off. If it enters the soil it may be stored and later taken up by plants, it may move into groundwaters or move laterally through the earth, appearing later in springs. This partitioning of rainfall determines whether a rainstorm results in a replenishing rain or a damaging flood. The movement of water through soils to streams, lakes, and groundwater is an essential component of the hydrological cycle.

The biological, chemical, and physical processes that occur in soils buffer environmental changes in air quality, water quality, and global climate. The soil matrix is the major incubation chamber for the decomposition of organic wastes, for example, pesticides, sewage, and solid wastes. Depending on how they are managed, soils can be important sources or sinks of carbon dioxide and other gases, also known as greenhouse gases, that contribute to the so-called greenhouse effect. Soils store, degrade, or immobilize nitrates, phosphorus, pesticides, and other substances that can become air or water pollutants.

Soil degradation through erosion, compaction. loss of biological activity, acidification, salinization, or other processes can reduce soil quality. These processes reduce soil quality by changing the soil attributes, such as nutrient status, organic and labile carbon content (organic carbon is the total amount of carbon held in the organic matter in the soil; labile carbon is that fraction of organic carbon that is most readily decomposable by soil microorganisms), texture, available water-holding capacity (the amount of water that can be held in the soil and made available to plants), structure, maximum rooting depth, and pH (a measure of the acidity or alkalinity). Some changes in these soil attributes can be reversed by external inputs. Nutrient losses, for example, can be replaced by adding fertilizers. Other changes such as loss of the soil depth available for rooting because of soil erosion or degradation of soil structure because of subsoil compaction are much more difficult to reverse.

Soil Quality and Agricultural Productivity

Damage to agricultural productivity has historically been the major concern regarding soil degradation. Agricultural technology has, in some cases, improved the quality of soils.

Agricultural management systems have been historically adopted without recognizing consequences on soil conservation and environmental quality, and therefore significant decline in soil quality has occurred worldwide. Soil quality has been defined as "the capacity of a specific kind of soil to function with its surroundings, sustain plant and animal productivity, maintain or enhance soil, water and air quality and support human health and habitation". Any evaluations of the quality of the soils must consider the multiple soil uses (e.g. agricultural production, forest, rangeland, recreation or urban development). However, the most widely accepted concept of soil quality and the most significant in a global context is referred to agro-ecosystems. Using soil quality concept along with identifying key attributes or indicators among the nearly infinite list of parameters, and methods for evaluating and monitoring it with respect to the numerous soil functions is an evolving process. Unlike water and air quality, simple standards for individual soil quality indicators do not appear to be sufficient because numerous interactions and trade-offs must be considered.

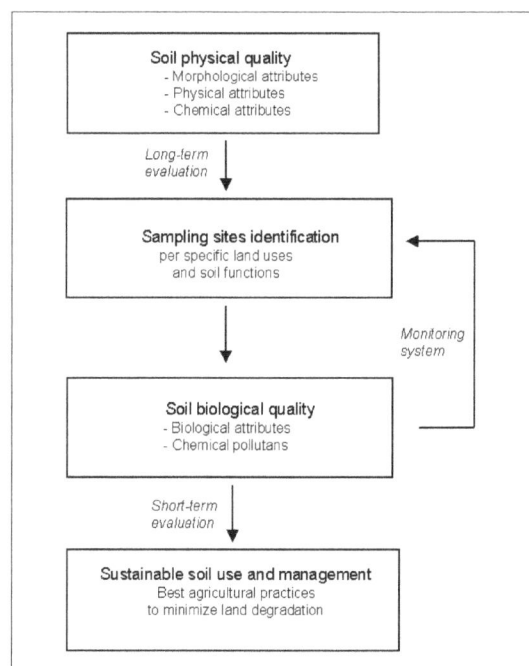

General approach to formulate sustainable soil use and management strategies
developing soil quality assessment on the basis of land evaluation analysis.

Soil quality evaluation or assessment tries to predict the natural ability of each soil to function, and land evaluation tries to predict land behavior for each particular use. However, land evaluation is not the same as soil quality assessment, basically because land refers not only to soil but to the combined resources of soil, terrain, water and land use. Also, the biological parameters of the soil are not considered by land evaluation. Soil physical/chemical indicators (e.g. soil-depth, texture, bulk density, infiltration, water retention, pH, carbonate content or cation exchange capacity), which are the main input soil characteristics in land evaluation, are very fix and permanent in the time. Soil biological indicators (e.g. populations of micro-organisms, microbial biomass and/or respiration, mycorrhizal association, nematode communities, enzymes or detailed characterization of organic matter) are very dynamic and exceptionally sensitive to changes in soil conditions. The biological attributes, along with the trace metals concentration, appear to be most sensitive to soil conservation/restoration and management practices such as crop rotation, conservation tillage and organic amendments.

For soil quality assessment, the development of relationships between all the soil quality indicators and the numerous soil functions may be a monumental task. Within this complex context, land evaluation models may serve as a first step to develop a soil quality assessment procedure. As shown in figure, land evaluation can be an appropriate procedure for analyzing the soil physical quality from the point of view of the long-term changes. Then, a short-term evaluation and monitoring procedure can be considered mainly for the soil biological quality and metal pollution of soil. Also, according to the complex nature of the soil, the high spatial and temporal variability of its different parameters, it appears appropriate to develop soil quality assessment based exclusively on biological indicators after the traditional land evaluation using basically physical parameters.

MicroLEIS, through its different land evaluation models, analyses the influence of selected physical indicators on critical soil functions referred to land productivity: agricultural and forest soil suitability, crop growth and natural fertility; and referred to land degradation: runoff and leaching potential, erosion resistance, pollutants absorption and mobility, and subsoil compaction. Therefore, this system appears to be an appropriate approach to develop the first soil physical quality evaluation. Since the lately 1980s, MicroLEIS has evolved significantly towards an agro-ecological decision support system on Internet. Presently, this system is a set of useful tools for decision-making, which can be applied to a wide range of land productivity and land degradation schemes. The design philosophy corresponds to a toolkit approach, where many software instruments are integrated: databases, statistics, expert systems, neural networks, Web and GIS applications and other information technologies.

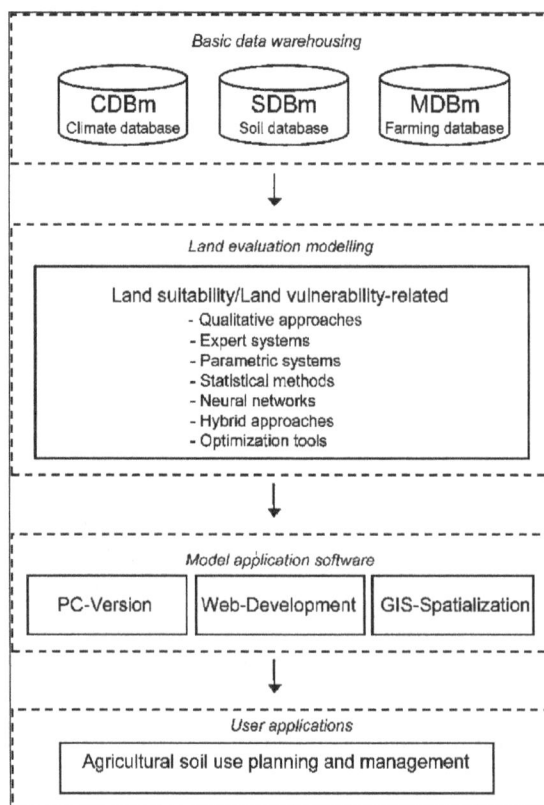

Conceptual design and components integration of the current status
of MicroLEIS agro-ecological decision support system.

Evaluating and monitoring soil quality is a complex undertaking. Knowledge-based decision support systems considering separately soil physical quality and soil biological quality appear to be an appropriate way to formulate, for each unique soil, the best agricultural practices or sustainable soil management systems to minimize land degradation processes such as soil erosion, compaction, salinization, contamination or loss of fertility.

References

- Environmental-monitoring- 5503: safeopedia.com, Retrieved 05 June, 2019

- Environmental-monitoring, earth-and-planetary-sciences: sciencedirect.com, Retrieved 18 July, 2019

- All-about-effective-air-quality-monitoring-46494: downtoearth.org.in, Retrieved 04 May, 2019

- What-is-air-quality: conserve-energy-future.com, Retrieved 23 April, 2019

- Water-quality-monitoring-115: adcon.com, Retrieved 24 August, 2019

- Why-is-water-quality-monitoring-important, breaking-news – 34104: envirotech-online.com, Retrieved 19 March, 2019

- Waterquality: floridakeys.noaa.gov, Retrieved 26 January, 2019

- Soil-quality-evaluation-and-monitoring- 242607437: researchgate.net, Retrieved 14 March, 2019

Permissions

Index

www.ingramcontent.com/pod-product-compliance
Lightning Source LLC
Chambersburg PA
CBHW082011190326

41458CB00010B/3155